The Genetics Explosion

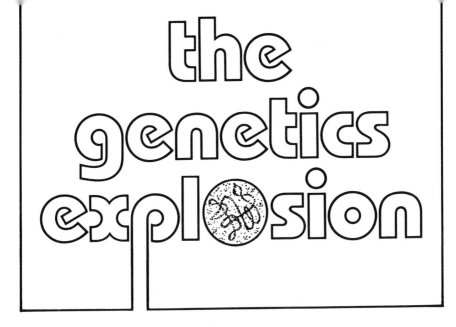

the genetics explosion

by Alvin and Virginia Silverstein

FOUR WINDS PRESS   NEW YORK

For David Reuther

LIBRARY OF CONGRESS CATALOGING IN PUBLICATION DATA

Silverstein, Alvin.
The genetics explosion.

Bibliography: p.
Includes index.
SUMMARY: Discusses genetics and genetic diseases
and explores the controversy over genetic
engineering with both its potential benefits and
its potential health hazards.
1. Genetic engineering—Juvenile literature.
[1. Genetic engineering.   2. Genetics]
I. Silverstein, Virginia B., joint author.
II. Title.   [DNLM: 1. Genetic intervention—
Popular works.   QH442 S587g]
QH442.S54   575.1   79-22651

ISBN 0-590-07517-9

Published by Four Winds Press
A Division of Scholastic Magazines, Inc., New York, N.Y.
Text copyright © 1980 by Alvin and Virginia Silverstein
Drawings copyright © 1980 by Scholastic Magazines, Inc.
All rights reserved
Printed in the United States of America
Library of Congress Catalog Card Number: 79-22651
Drawings by Constance Ftera and Richard Erik Warren
Book design by Constance Ftera
1 2 3 4 5 84 83 82 81 80

PHOTO CREDITS

Brookhaven National Laboratory, 6, 25
General Electric, 141
Constance Ftera, 14, 19, 20, 23, 28, 29, 31, 36, 70, 86, 102, 108, 126, 128
National Institutes of Health, 62 (both), 95 (top)
United Press International, 5, 95 (bottom)
Richard Erik Warren, 33, 44, 55, 75, 99, 113, 120, 132
Wide World, 3

Contents

The Genetics Explosion

If you listen carefully, you may hear an explosion. It's not the kind that goes BANG! No, it's a different kind—a knowledge explosion. Every day thousands of new facts pour out of laboratories throughout the world. Each day we learn more about the world around us and about ourselves.

Over the next two years, scientists will gain more knowledge than people did in all of human history before you were born. Imagine that! In just two years the mountain of new knowledge will be bigger than the mountain that was built up over many thousands of years. We can already see the results of this knowledge explosion. For example, your pocket calculator is more powerful than the million-dollar computers of thirty years ago.

1

Things are happening so fast in science these days that it is hard for people to keep up. But there is an important area of science that is moving especially fast: molecular genetics—the study of the chemicals of heredity.

You have probably noticed that you look rather like your parents. Perhaps you have eyes like your mother's or a nose that looks just like your father's. Turning the pages of the family photo album, you can trace resemblances to other relatives—perhaps you inherited your grandfather's chin, or hair just the color of your aunt's when she was young. All these traits have been passed down to you, combined into a new pattern that is uniquely yours.

You started life as a single round cell, too small to be seen without a microscope. Yet that single cell contained a complete set of blueprints for building a person and a detailed instruction manual for day-to-day operation. All this information was neatly tucked away inside 46 tiny rod-shaped structures called chromosomes and spelled out chemically in a substance called *deoxyribonucleic acid* (DNA for short). Portions of this DNA, called genes, contain the information needed to produce hormones, enzymes, and other important body chemicals, and to determine traits such as blue eyes, brown hair, and a turned-up nose.

The single cell that was the start of you carefully duplicated its hereditary information, chemical by chemical. Then it divided to form two identical cells, each with a complete set of your own, unique DNA molecules. The two new cells divided in turn, again and again. Gradually, different-looking cells appeared and

DNA molecules are shown about 30,000 times their normal size in this electron micrograph taken by Prof. David Jackson of the University of Michigan.

formed the different parts of your body—brain, heart, eyes, skin, muscles, and so on. Yet before dividing, each cell made its own careful copy of your hereditary information. So even now, each of your trillions of cells has its own complete set of instructions for building your whole body!

Only a small fraction of the DNA in any particular cell is actually working, translating particular instructions into materials to be used for building or maintaining the body. This working DNA contains the instructions needed for that cell's own jobs. The rest of the cell's DNA is "turned off." It is there, and if the cell divides, all of the DNA will be faithfully duplicated and passed on to the new "daughter cells." But most of the DNA isn't doing anything. Imagine the mix-ups that would occur if this were not the case. You might find yourself growing toenails on your eyelids, or eyes on your elbows!

Ideally, each person comes into the world with a set of chemical blueprints for building a strong, healthy body and keeping it that way throughout a long life. But mistakes can happen. Sometimes errors occur in copying the cell's DNA. This doesn't happen very often, but think of how many times cells must divide in order to go from the single cell that starts off a baby to the trillions of cells in the body of an adult. DNA may also be altered by radiations or chemicals. With so many opportunities for changes, it's not surprising that scientists estimate we each are carrying from three to eight *mutations,* or changes, in our genes. Many of these mutations have no apparent effect on us, but there are more than two thousand different genetic diseases known to be the result of something wrong with the chromosomes.

Disease and accidents can also damage the body, leading to illness, disability, and death. If a person loses an arm in an auto accident or has a lung destroyed by cancer, theoretically the healthy cells that remain have all the blueprints needed to build a replacement arm or lung. But they don't do it. The genes for building new arms and lungs are turned off very early in a person's development—long before birth—after the organs of the body have been formed. In most cases, they are not turned on again.

Wouldn't it be wonderful if we could fix damaged DNA, so that people with genetic diseases could be cured? And if we knew how to turn genes on and off when we wanted to, people could grow their own

This model of a single gene, more than a quarter of a million times life size, was constructed by the Upjohn Company, Kalamazoo, Michigan.

A Brookhaven National Laboratory researcher checks a refrigerator full of cultures of mutant forms of the bacteriophage T7, used in genetic studies.

replacement parts! We can't do either of these things yet. Not too many years ago, the idea that we would *ever* be able to do them seemed to be merely wishful thinking, or "science fiction." But now information about genetics—down to the most fundamental level— is flooding in. We have already learned an enormous amount about what the genes are and how they work, and each day we learn more. Now scientists are beginning to work out practical applications of this genetic kowledge. We have progressed from *genetics,* the science that describes heredity, to *genetic engineering,* the application of genetic knowledge for practical purposes. We are gaining the ability to change the living creatures of our world—and to remake ourselves.

What benefits do scientists speculate that the new genetic knowledge may bring? A quarter of a million American children are born each year with birth defects, which may leave them crippled, disfigured, or mentally retarded. Most of these defects have genetic causes, and with enough knowledge we might be able to prevent them. Errors in genes can also cause metabolic diseases, in which chemical reactions in the body don't work properly. Thousands of chemical reactions go on constantly in our bodies, and they are closely in- terlinked, each reaction depending on many others. These metabolic disorders can cause illness, retarded development, and death. If we had effective ways of locating the faulty genes and correcting them, we could cure the diseases. We might also be able to build in resistance to infectious diseases caused by bacteria and viruses, and help people to repair their own damaged body parts. Genetic engineers are working on other creatures, too, breeding new varieties of plants, for example, which would grow faster and produce larger crops of more nutritious food. Microorganisms—which are already being grown in huge factories to produce a variety of products, from fermented beverages to antibiotics—could be made to order, giving us vast new supplies of hormones, enzymes, and other important biochemicals.

Some people believe that there is another side to the genetic engineering story—that the new knowledge and techniques scientists are developing may bring not only benefits, but also risks. They fear that the new knowledge may be misused—perhaps on purpose, perhaps accidentally—if we try to do too much, before we know how to do it properly. Recently one area of genetic engineering, recombinant DNA research, has stirred a furious debate. In recombinant DNA techniques, genes from other organisms—often plants and animals—are inserted into bacteria. Some scientists and citizens feared that we might accidentally start a horrifying plague that could wipe out the defenseless inhabitants of our planet. As more has been learned about tinkering with bacterial genes, most of these fears have died down. But other sensational news in areas related to the new genetics research, from the birth of the first "test-tube baby" to the publication of a book that claimed to describe the cloning of a human being, has continued to focus attention on some of the deeper questions that this new research is raising. Do we have the right to tamper with our genes and those of other creatures? Might we seriously unbalance the community of life on our planet which has developed over millions of years of gradual change and evolution? Are we "playing God" when we consider the prospect of remaking ourselves?

The recombinant DNA debate brought the average citizen into a far more active role in scientific decision making than ever before. In Cambridge, Massachusetts, and in other communities, people were suddenly faced with the task of deciding whether it was safe and wise to allow certain new types of genetics research to be conducted—decisions that might ultimately affect the whole world's history. It is likely that this kind of debate will continue as the exploding field of genetics research continues to bring stunning new advances. Now more than ever, we all need to learn more about this new science so that our decisions will be responsible ones.

ABCs of Genetics

We can see the work of a kind of genetic engineering all around us: sleek racehorses; the varied breeds of dogs, from tiny Chihuahuas to huge St. Bernards; and the corn, wheat, and other grains, vegetables, and fruits we eat. For thousands of years, people have been breeding animals and plants, selecting individuals with valuable traits and trying to improve them and pass their qualities on to future generations. Shrewd breeders gained a great deal of practical knowledge of heredity. But until a little more than a century ago, no one had worked out a coherent scientific theory of genetics. And it was even later that scientists began to realize what caused the great variety among individual animals and plants.

The modern age of genetics began in the mid-eighteen

hundreds. In a monastery in the town of Brünn in Austria (now Brno, Czechoslovakia), a monk named Gregor Mendel was wondering about heredity. Why was it that some animals and plants "bred true"—all their offspring looked just like the parents, for generation after generation—while others that *seemed* just like them had offspring with a variety of traits often unlike their parents'? Mendel puzzled over the peas growing in the monastery garden. Usually the seeds from tall pea plants produced more tall plants; but sometimes the new plants were dwarfs. Plants grown from yellow seeds often had yellow seeds of their own, but sometimes their seeds were green. And so it was for other traits as well, such as the color of the flowers, smoothness or wrinkling of seeds, and so forth.

Gregor Mendel was not the first person to wonder about such matters. But he was unusual in that he searched for the answers to his questions in a very practical way. Over a period of more than ten years, the monk grew thousands of pea plants in his quiet monastery garden. First he carefully bred a number of pure lines: plants that always produced yellow pods and plants that always had green ones; plants with yellow seeds and plants with green seeds; plants with red flowers and plants with white ones; tall plants and dwarf plants; and so on. When he was ready to go on to the next part of his experiment, he had seven pairs of pure lines, each giving pea plants with alternative variations of a particular trait.

Then Mendel began crossing the pairs of lines, carefully snipping off the stamens (the male parts) of one flower and dusting its pistil (the female organ) with pollen from another. When seeds formed, he planted them, and when the plants of the new generation were ready to be pollinated, he carefully covered the flowers so that each would be pollinated with its own pollen. As pea generation followed generation, Gregor Mendel continued his experiments, sometimes cross-pollinating plants of two different types and sometimes self-pollinating them. He painstakingly tallied the results of each test, counting hundreds and thousands of plants, flowers, pods, and seeds, and writing all the totals down in his notebooks.

Another biologist, looking at the results of Mendel's experiments, might have seen only a hopeless jumble of numbers. But Gregor Mendel was a rather mathematical-minded person, and for him those numbers soon began to form interesting patterns. In one experiment, for example, he crossed peas that produced green seeds with peas that produced yellow seeds. The cross produced plants with a total of 258 seeds, and all were yellow. Mendel planted those yellow seeds and carefully self-pollinated them. They produced a total of 8,023 seeds: 6,022 yellow and 2,001 green. There were about three yellow seeds for every green one. But Mendel didn't stop there. He planted the second-generation seeds and self-pollinated the plants. Now the result was somewhat different. The plants grown from the green seeds produced only green seeds. These plants had bred true for green seeds, even though their parents had had yellow seeds. About a third of the second-generation yellow seeds grew into plants that produced only yellow seeds: These plants also bred true. The rest of the yellow seeds produced plants with the same three-to-one ratio of yellow to green seeds. Apparently the three-to-one ratio was a little more complicated than it had first seemed. For every yellow seed that bred true there were two yellow seeds that gave mixed offspring and one green seed that bred true.

What did all this mean? Gregor Mendel was only an amateur biologist who had never even managed to pass the qualifying exam for certification as a natural history teacher. But now he proved that he not only was a careful experimenter, but also had a touch of genius. With a flash of insight he found the answer. Each trait, he thought, is inherited in distinct units, or factors. Each of his pea plants had two of these factors for each trait; one came from the mother plant and the other from the father. In mating (pollination of the plants) the factors separated, with only one from each parent being passed on to a particular offspring; but the offspring, receiving one factor from each parent, again possessed the full set of two factors for each trait. Some factors seemed to be dominant over others: Whenever they were present, they determined the trait. The other factor of the pair, called the recessive, seemed to shrink away into hiding. But its effects might appear again in the next generation, if it happened to combine with a similar recessive factor.

In the pea experiment described, for example, the original pure lines could be thought of as YY (producing only yellow seeds) and gg (all green seeds). When the lines are crossed, all the seeds produced have a Y factor from one parent and a g factor from the other. So their genetic makeup—their *genotype*—is Yg. These seeds are all yellow because Y is dominant. Then what happens

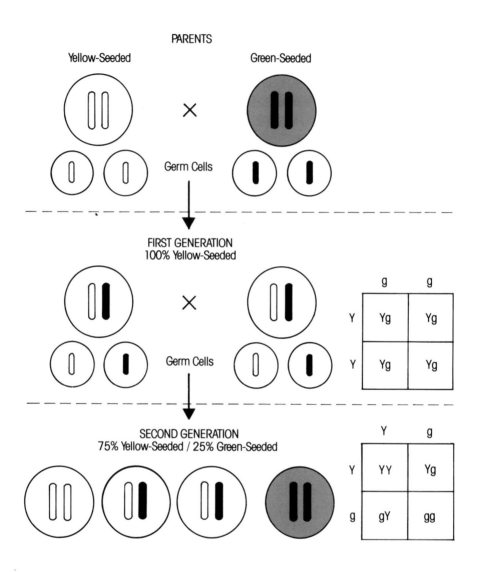

One of Mendel's experiments: in a cross of green-seeded peas with yellow-seeded peas, the yellow-seed trait is found to be dominant. The small tabulations of the possible results of the crosses at the right are called Punnett squares.

when the plants grown from Yg seeds are crossed with each other or self-pollinated? The mother is Yg and the father is Yg. About a quarter of the seeds of the next generation will receive a Y factor from each parent and will be YY. The plants grown from these seeds will have all yellow seeds and will breed true. About a quarter of the seeds will receive a g factor from each parent. They will have a gg genotype. Their *phenotype*—the outward appearance of the trait—will be all green seeds, and they, too, will breed true. The rest of the seeds (about half) will receive a Y factor from one parent and a g factor from the other. Their genotype will be Yg, and their phenotype will be yellow seeds; but like the first-generation seeds, they will not breed true. A mixed genotype, such as Yg, is called *heterozygous,* while a pure genotype, such as YY or gg, is called *homozygous.*

Mendel's theories agree beautifully with what we know about genetics today. His "factors" correspond to the genes on the chromosomes. Each animal or plant has two of each gene, because the complete chromosome set can be grouped into a series of corresponding pairs. (The pea actually turns out to have seven pairs of chromosomes, one corresponding to each pair of traits that Gregor Mendel studied.) When the germ cells—the special sex cells that will join to form new organisms—are being produced, a special kind of cell division occurs. Instead of receiving a complete copy of the cell's DNA, each daughter cell receives only half of the chromosome set, one of each pair.

Today all this seems obvious. But in Mendel's time, no one had ever heard of chromosomes or genes. Mendel sent a copy of his ideas to one of the most famous biologists of the time; but the scientist didn't understand the monk's careful calculations and dismissed him with a short note, cold and polite. Mendel was crushed. He was just an amateur after all, and his dreams of publishing his work in a great scientific journal were too presumptuous. In 1865 he gave a talk about his experiments and theories at the local natural history society in Brünn, and over the next few years he published several papers in the little journal published by the society. Mendel was elected abbot of his monastery and lived a contented life for two more decades, too busy with his duties to spend much time on experiments. Meanwhile, his papers gathered dust in the library. No one remembered them, and Mendel never knew that he would one day be famous as the "Father of Genetics."

A generation went by. Then, suddenly, an extraordinary coincidence occurred. In 1900 three different scientists, working separately—Hugo de Vries in the Netherlands, Karl Correns in Germany, and Erich von Tschermak in Austria—discovered the basic principles of heredity. Conscientious scientists, they made a careful search of the scientific literature before writing up their own findings. All three of them discovered Mendel's papers and realized that their discovery was actually a rediscovery. One of the most exciting experiences for a scientist is to be hailed as the discoverer of an important new principle. It must have been a terrible temptation for each of these researchers to say nothing about those old, forgotten journals of the Brünn society and announce the laws he had worked out as his own. But in a striking example of honesty and high principles, each scientist reported his own findings only as supporting evidence, giving full credit to Gregor Mendel as the true pioneer in genetics.

The dramatic triple rediscovery of Mendel's work excited biologists all over the world. Eagerly they plunged into work in the new field of genetics. They soon discovered a number of exceptions and additions to Mendel's laws of heredity. The Dutch botanist Hugo de Vries studied cases in which sudden changes in traits occurred among the offspring of animals and plants, which did not seem to be explainable by the normal hereditary laws, and were then passed on to further generations. He coined the term *mutation* for these hereditary changes.

In the United States, geneticist Thomas Hunt Morgan made a particularly fortunate selection of an experimental animal: the fruit fly, *Drosophila melanogaster*. Mendel needed a whole field to grow his thousands of peas, and when he gathered the seeds he often had to wait until the following year to grow the next generation. Fruit flies, on the other hand, can be rapidly bred by the hundreds in bottles on a laboratory shelf. It is easy to tell males from females, so the experimenter can conveniently arrange suitable matings. The female lays several hundred tiny eggs, which develop within ten days or so into adult flies ready to mate and start a new generation on its way.

Fruit flies show a great deal of natural variation. Most have a gray body, but some have a body that is black or some other color. The usual red eyes may vary to shades of brown, amber, or even white. Sometimes fruit flies are born with tiny "vestigial" wings, which cannot be used for flight. Morgan studied the inheritance of these variations in his fruit flies and discovered some important exceptions to Mendel's laws. Mendel had said that traits are inherited independently. For example, a pea plant with white flowers might have either yellow seeds or green seeds, and the seeds might be either wrinkled or smooth. This independent inheritance was true for the seven traits that Mendel studied in his peas, but Morgan found a different pattern in his fruit flies. Some combinations of traits seemed to be passed along together. For example, if a pure-bred fly with a gray body and normal wings was mated with a black-bodied, vestigial-winged fly, all the offspring had gray bodies and normal wings. But their children—the second generation—included about three gray-bodied, normal-winged flies to every one that had a black body and vestigial wings. There were practically no flies with gray bodies and vestigial wings, or with black bodies and normal wings. Morgan concluded that these traits seemed to be linked together.

Some traits seemed to be linked to the flies' sex. In one of his broods of flies, Morgan noticed a male with white eyes. He mated it with a normal red-eyed female and obtained a curious result. The first-generation offspring were all red-eyed—so the new white-eye mutation was recessive. When these flies were bred together, the expected 3:1 ratio of red eyes to white ones appeared in the second generation. But then Morgan looked closer. *All* the females had red eyes. Half of the males had red eyes, and half had white eyes. There were no white-eyed females in that generation at all. But white-eyed females were obtained in the next

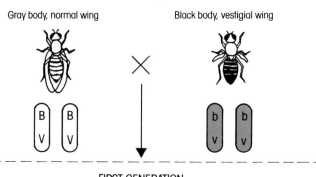

PARENTS

Gray body, normal wing

Black body, vestigial wing

FIRST GENERATION
100% gray body, normal wing

SECOND GENERATION

Nearly 75% gray body, normal wing

Nearly 75% black body, vestigial wing

And a few gray body, vestigial wing, and black body, normal wing

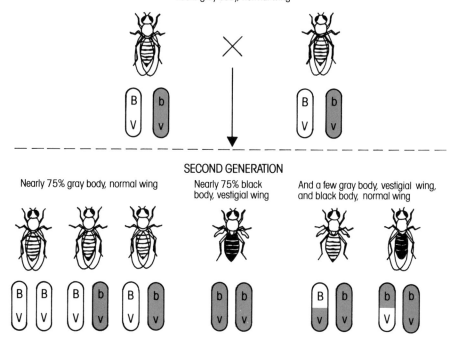

One of Morgan's experiments: some traits, such as body color and wing shape in the fruit fly, are not inherited separately, but seem to be linked together. Occasional cases of separate inheritance can be explained by crossing-over (see pp. 22–24).

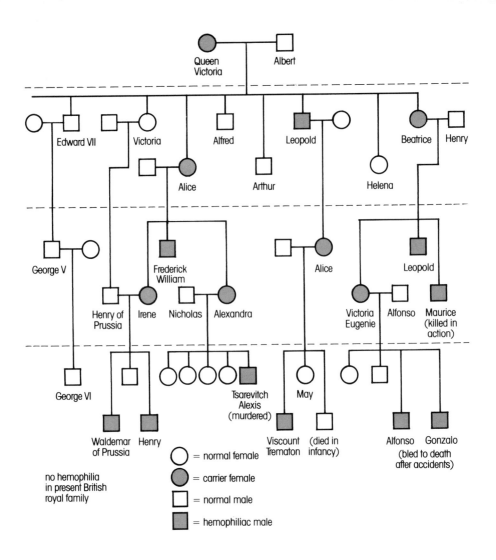

The hemophilia pedigree among the descendants of Queen Victoria. (Many of the grandchildren and great-grandchildren who did not inherit the hemophilia gene have been omitted.)

generation, when white-eyed males were mated with the heterozygous (mixed-genotype) females.

Sex-linked traits have also been observed in humans. Color blindness occurs most often in men, and only very rarely in women. Hemophilia, the "bleeder's disease" in which blood does not clot properly after a cut or wound, is also found mainly in men. Some famous cases of hemophilia occurred in the royal families of Europe during the past century, and a detailed pedigree has been drawn up, tracing the inheritance of this sex-linked trait. These cases apparently originated with Queen Victoria. She herself did not have hemophilia, but she was carrying the trait, for one of her sons suffered from the disease. Her other seven children didn't *seem* to be affected, but two of her daughters turned out to have inherited a hemophilia gene. Cases of the disease turned up among their sons and grandsons. One of the stricken boys was the Tsarevitch Alexis of Russia. This beloved only son was sickly and constantly in danger of bleeding to death. His anxious parents sought everywhere for a cure. Eventually they fell under the influence of Rasputin, who not only advised them on the care of their son, but came to share some of their power in ruling the kingdom. Rasputin's influence on the czar and czarina is thought to have been one of the indirect causes of the Russian Revolution. So in this case, a sex-linked genetic disease may have changed the history of the world.

Gradually, further insights were gained into the nature of heredity. In 1879 German biologist Walter Flemming discovered that when he treated cells from the tail fin of a salamander with a dye, tiny threadlike structures inside the nucleus of each cell took up the color and became clearly visible under a microscope. These structures were named *chromosomes,* from the Greek words meaning "color bodies." In the early 1900s W.S. Sutton proposed that these chromosomes were the determiners of heredity. Morgan suggested the term *gene* for the portion of a chromosome determining a particular trait. This new theory of chromosomes and genes could neatly explain his own findings, as well as those of Mendel.

Picture a chromosome as a string of beads, each bead representing a gene. Just before a cell divides to produce germ cells, the chromosomes pair up and then separate, with one chromosome of each parent pair going to each of the new cells. If two genes happened to be on the same chromosome, they both would be transferred to the same germ cell and so would be inherited together. So there is an actual physical linkage of genes on the chromosomes to explain why we inherit linked traits.

The idea of genes on a chromosome also explains another of Morgan's important findings. In linkage he had discovered an exception to one of Mendel's laws, but he also discovered an exception to the exception. Remember we said, "There were *practically* no flies with gray bodies and vestigial wings, or black bodies and normal wings." But there were a few. Morgan theorized that when the chromosomes are pairing up before division, they may get tangled. When they pull apart and separate, parts of the chromosomes may break and then join back together. Sometimes, in this rejoining, the end of one chromosome might get attached to the other, and vice versa. Afterward there

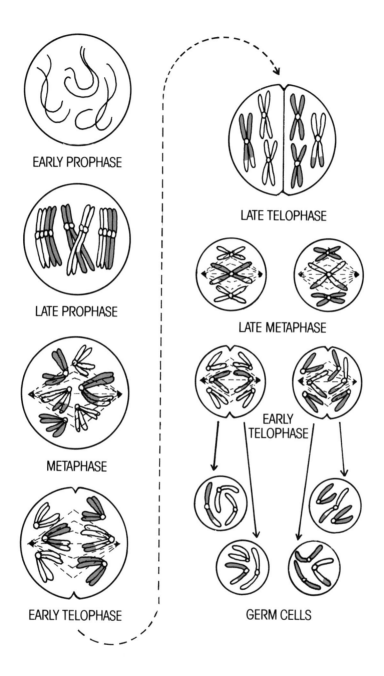

EARLY PROPHASE

LATE PROPHASE

METAPHASE

EARLY TELOPHASE

LATE TELOPHASE

LATE METAPHASE

EARLY TELOPHASE

GERM CELLS

Meiosis, the special kind of cell division that produces the germ cells.

would still be two chromosomes with about the right number of genes, but not all of them would be the genes they started out with. Morgan named this kind of genetic accident *crossing-over,* and in the years since he proposed his theory, biologists have actually observed chromosomes doing just what he predicted.

An important conclusion from Morgan's crossing-over theory is that the closer together genes are positioned on the chromosomes, the more likely they are to stay linked. By calculating the percentage of cases in which combinations of traits are inherited together and the percentage in which they are passed on independently, scientists have been able to draw up gene maps, locating the genes for a number of traits on the chromosomes of fruit flies, mice, and even humans.

Another boost to genetics research was provided in the late 1920s, when American biologist Herman J. Muller discovered that X rays could produce mutations in fruit flies. Other types of radiation (including the ultraviolet rays of the sun) and various chemicals were later found to have similar *mutagenic* effects. Now researchers could produce large numbers of varied mutations to work with, without having to wait through many thousands of matings for the rare occurrence of natural mutations.

Real progress in learning about heredity at the chemical level came after researchers had begun to shift their attention to another kind of experimental organism: the microorganism. Bread molds, yeasts, and bacteria can easily be grown in the laboratory in test tubes or shallow culture dishes. In a single dish a researcher can study not hundreds or thousands of

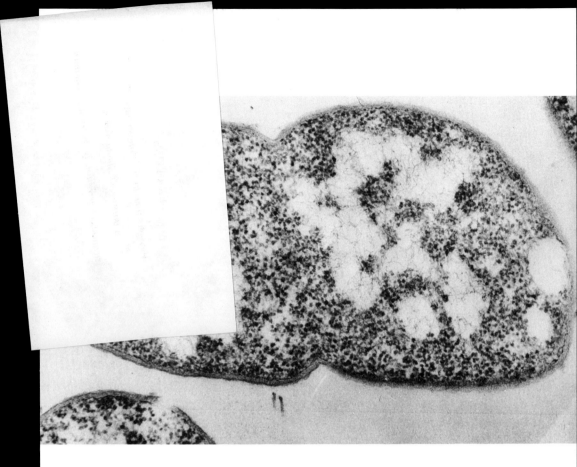

A thin section of a dividing cell of E. coli, *a bacterium commonly used in laboratory experiments.*

organisms, but millions of them. The common bacterium *Escherichia coli* (better known as *E. coli*) has its natural habitat in human intestines but has adapted nicely to life in a culture dish. With plenty of food and just the right amounts of moisture and warmth, an *E. coli* cell can pass through an entire generation in just twenty minutes, dividing to form two miniature copies of itself. Theoretically, under ideal conditions such a bacterium could produce, in just one day, more than 1,000,000,000,000,000,000,000 descendants, with a total weight of more than 200 tons! Of course, real conditions are not anywhere near that ideal, but in the usual laboratory setting a single microscopic bacterium can fairly quickly produce a good-sized colony, easily visible to the naked eye.

A group of cells, all produced from an original single-celled ancestor, is referred to as a *clone*. (In a way, you, too, might be thought of as a clone, since your entire body, containing trillions of cells, was all derived from a single original cell, and all your cells share the same heredity.) Though bacteria usually reproduce in an *asexual* manner, with a single parent producing two identical daughter cells, they sometimes engage in a kind of mating called *conjugation,* in which two bacteria can exchange some of their genetic information. Like peas, fruit flies, and people, bacteria and other microorganisms have a number of traits which may vary, and the variations show typical patterns of inheritance. Indeed, bacteria have chromosomes, and they contain the same kinds of chemicals as those of animals and plants.

Curiously, the chemical of heredity was discovered only a few years after Mendel read his paper to natural history buffs at Brünn. In 1869 a German chemist, Friedrich Miescher, isolated a chemical from the nuclei of cells and named it *nucleoprotein.* No one realized its significance at the time. As chemical techniques improved, it was found that the nucleoprotein is made up of a combination of protein and a substance called *nucleic acid.* Gradually it was learned that there are two main types of nucleic acid: deoxyribonucleic acid (DNA) and ribonucleic acid (RNA). The nucleic acid in the nucleus is mainly DNA, while most of the RNA is found in other parts of the cell. Reactions with dyes revealed that the chromosomes are made up of nucleoprotein, and thus it seemed that this was the chemical of heredity. But which part of the nucleoprotein?

Some evidence seemed to indicate that nucleic acids

the experts. Enormous amounts of information must be needed, they reasoned, to produce a complicated chemical like the hemoglobin in blood or to specify that eyes will be brown and not blue. Nucleic acids are large molecules, but the building blocks that form them are of only four kinds. The chemical units of the nucleic acids are called *nucleotides,* and they consist of three parts: a phosphate, a sugar, and a substance called a nitrogen base. Four kinds of nitrogen bases are found in DNA: adenine, thymine, guanine, and cytosine (usually abbreviated as A, T, G, and C). RNA also has four kinds of bases: Three are the same as DNA's—A, G, and C—but another base, uracil (U), takes the place of thymine (T). What kind of message could you spell out with only four letters? (The last sentence alone used twenty-one different letters.) Think about it, though. Morse code has only two letters, "dot" and "dash." Yet any message in the English language can be spelled out in Morse code. Computers work on a number system with just two numbers, "plus" and "minus," yet they can do calculations involving numbers larger than you could even imagine. Apparently an "alphabet" of only four kinds of nucleotides could spell out complicated messages after all—complicated enough to contain the information of heredity.

Eventually the puzzle came clear. In 1928 an English researcher, Frederick Griffith, had discovered something he called the "transforming principle." He worked with two strains of *Pneumococcus,* the pneumonia bacterium. One strain was very deadly— when it was injected into rats it killed them. The other strain was rather weak, and an injection of it did not

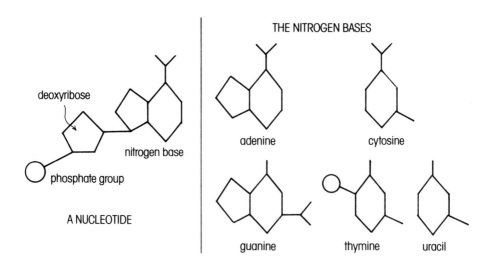

The building blocks of the nucleic acids are nucleotides, which contain a sugar, a phosphate group, and a nitrogen base.

harm the rats. If the deadly strain of the bacterium was killed by heating, it, too, was harmless. But Griffith found that if he injected a mixture of killed bacteria of the deadly strain and live bacteria of the harmless strain, the rats died. Apparently something in the killed bacteria was transforming the harmless bacteria into deadly ones.

In the early 1940s the Canadian research team of O.T. Avery, C.M. McLeod, and M. McCarty repeated Griffith's experiments, with some new refinements. They separated various chemical components of the killed bacteria and tried mixing each one with the live harmless strain. Only the DNA fraction could transform the harmless bacteria into deadly forms. DNA was the transforming principle.

Attention was now focused on DNA, and researchers sought to discover how the parts of its molecule are put together and how it works. There were no microscopes

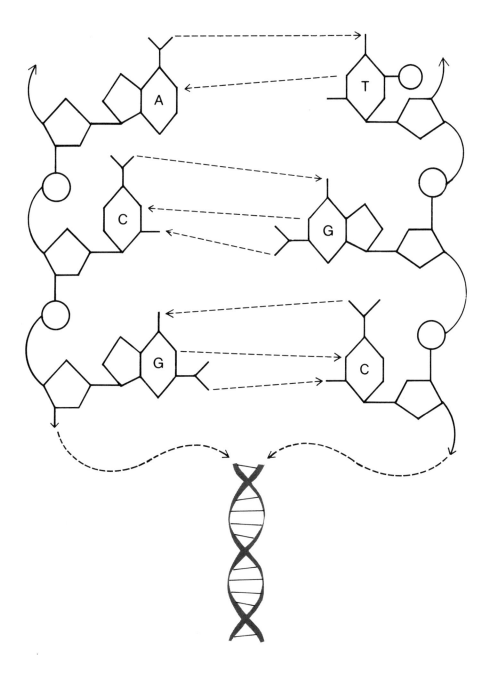

The Watson-Crick model of DNA. Nucleotides are linked together into a double helix.

powerful enough to "see" a DNA molecule. But scientists did have an indirect approach, called X-ray diffraction study. They bounced beams of X rays off various parts of the molecule and then analyzed the patterns formed by the bouncing X rays. Several research teams in various parts of the world were working on the problem. The race to work out the DNA structure was won in 1953 by a British biochemist, Francis Crick, and his young American colleague, James Watson.

The model of DNA that Watson and Crick proposed was elegantly simple. It not only neatly matched the X-ray data, but it also suggested insights into how DNA works in the cell. According to this model, DNA is pictured as a double helix, a corkscrew shape much like a spiral staircase. Twin backbones, each formed from sugar and phosphate, are coiled like springs. The nitrogen bases, attached to them, project inward and are fastened together by chemical bonds called hydrogen bonds, to form the "treads" of the staircase. The pairing of the bases on opposite chains of the double helix follows very definite chemical rules. A on one chain is always paired with T on the other, and C is always paired with G. Thus, if the bases on a portion of one chain spelled out AAGTCC, the corresponding portion of the other chain would contain the bases TTCAGG.

The Watson-Crick model was eventually confirmed by many new experimental findings. Indeed, when microscopes powerful enough to show a single DNA molecule were finally built, it turned out to look just as they had predicted. Meanwhile, with this structure to work from, Watson, Crick, and other biochemists soon

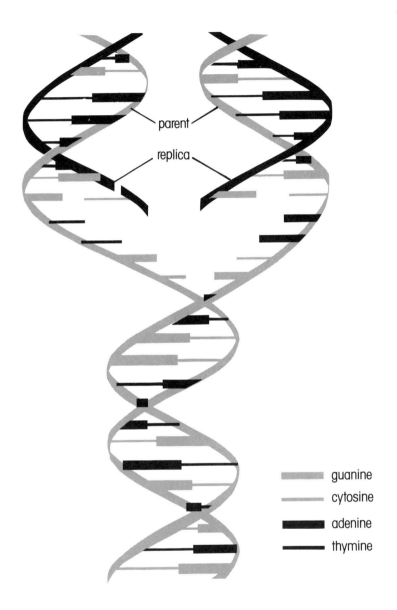

parent

replica

guanine
cytosine
adenine
thymine

DNA replication according to the Watson-Crick model. The two coiled strands of DNA unzip, and nucleotides are assembled on the uncovered templates, forming two "daughter" helices. Each will contain one strand of "old" DNA and one newly constructed strand.

fitted other pieces of the puzzle into place. RNA also has a helical structure, but it is a single helix, not a double one.

When a cell is getting ready to divide, the strands of its DNA begin to separate, like a zipper unzipping. As the separated strands are exposed, nucleotides line up opposite each one, all in the proper order for correct pairing of the bases—A with T and C with G. They join together into chains, and as the original DNA molecule continues to unwind, two complete new DNA molecules are gradually built up. When the process is complete, each new DNA molecule has one "old" strand and one newly assembled one. The process sounds very neat on paper, but it is rather extraordinary when you try to imagine it occurring in a tiny microscopic cell. Picture a typical *E. coli* cell, for example. Its DNA is coiled into about 360,000 turns. Each of these turns must unwind, and exactly the right nucleotides must be brought into just the right positions and joined together, then coiled up again into new double helices—and all within a few minutes, for within twenty minutes the bacterium will be ready to divide again.

The process by which DNA reproduces itself is referred to as *replication.* In a somewhat similar way, DNA can also serve as a *template,* or pattern, for the formation of RNA. When a portion of the DNA

In the nucleus, the DNA of a gene is transcribed into a messenger RNA copy. mRNA then goes out into the cytoplasm, and at the ribosomes it is translated into a protein, assembled according to the triplet code spelled out in RNA nucleotides from amino acids delivered by transfer RNA molecules.

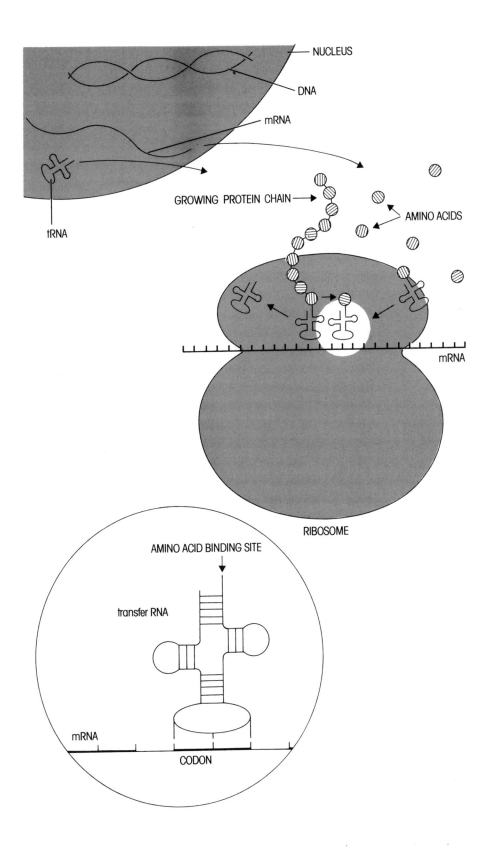

NUCLEUS

DNA

mRNA

GROWING PROTEIN CHAIN

AMINO ACIDS

tRNA

mRNA

RIBOSOME

AMINO ACID BINDING SITE

transfer RNA

mRNA

CODON

molecule—a gene—is exposed, RNA nucleotides pair up with the DNA bases and are assembled into a chain. Again, C pairs with G, but A pairs with U when RNA is being formed.

Studies revealed that there is more than one kind of RNA. A form called *messenger RNA* carries the "message" of the DNA gene out into the cell. Smaller RNA molecules called *transfer RNA* have the job of collecting amino acids to be built into proteins. Each kind of transfer RNA picks up only one particular kind of amino acid. Still another kind of RNA, *ribosomal RNA,* is found in cell structures called ribosomes, where proteins are produced according to the plan spelled out by messenger RNA.

How can a "message" spelled out in nitrogen bases in RNA dictate the exact structure of a protein? Researchers have discovered that the RNA bases form a code, in which groups of three "letters" correspond to specific amino acids, the building blocks of proteins. An alphabet of four letters can provide sixty-four different triplet combinations—far more than would be needed to correspond to the twenty or so amino acids found in proteins. So in the code of life, most of the amino acids are actually specified by more than one triplet, or *codon,* in messenger RNA. A dictionary of the genetic code has been worked out, so that if scientists know the structure of a gene, or of the messenger RNA formed by it, they can figure out the amino acid sequence of the protein that RNA will produce. The dictionary can also be used in the opposite direction: If researchers have worked out the structure of a protein, they can design a messenger RNA to produce it. (Because more than one codon may correspond to an amino acid, the artificial

RNA may not be exactly like the one a living cell uses; but it will work.)

DICTIONARY OF THE GENETIC CODE

AAU AAC	Asparagine	GAU GAC	Aspartic acid	CAU CAC	Histidine	UAU UAC	Tyrosine
AAA AAG	Lysine	GAA GAG	Glutamic acid	CAA CAG	Glutamine	UAA UAG	Stop signals
AGU AGC	Serine	GGU GGC	Glycine	CGU CGC	Arginine	UGU UGC	Cysteine
AGA AGG	Arginine	GGA GGG		CGA CGG		UGA UGG	Stop signal Tryptophan
AUU AUC	Isoleucine	GUU GUC	Valine	CUU CUC	Leucine	UUU UUC	Phenylalanine
AUA		GUA		CUA		UUA	Leucine
AUG	Methionine	GUG		CUG		UUG	
ACU ACC	Threonine	GCU GCC	Alanine	CCU CCC	Proline	UCU UCC	Serine
ACA ACG		GCA GCG		CCA CCG		UCA UCG	

Most of this information about the chemicals of heredity was obtained in experiments with bacteria (especially *E. coli*) and with even smaller and simpler microorganisms—viruses. Scientists still do not agree on whether viruses should be considered truly alive. Viruses can actually be crystallized, as can salt and other nonliving chemicals. They can be chemically taken apart and put back together again. But they do have one characteristic that seems typical of "life"—they can reproduce. They can also mutate and pass on their changed heredity to their offspring.

If a virus is indeed alive, it is certainly the ultimate, streamlined, stripped-down version of life. It consists of

LIFE CYCLE OF A BACTERIOPHAGE

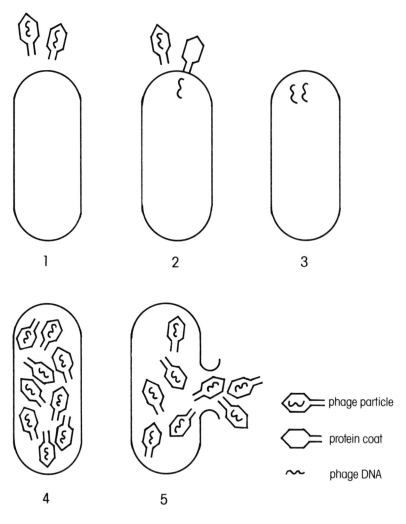

phage particle

protein coat

phage DNA

A virus cannot reproduce outside a living cell (1). It injects its nucleic acid into the cell it infects (2), then directs the cell to make virus nucleic acid and proteins (3), which are assembled into new virus particles (4). Finally the infected cell may burst (5), spilling out viruses ready to infect other cells.

a core of nucleic acid (DNA for some viruses, RNA for others), wrapped in a coat of protein. A virus can't reproduce all by itself. It must infect a living cell. Generally it sheds its protein coat, and its nucleic acid enters the cell. Once inside, the virus takes over and starts giving the cell orders. It provides the blueprints for making new viruses—hundreds of them. When the new protein coats and nucleic acid cores are assembled into viruses, the host cell bursts open and dies, spilling out the viruses, which can then infect other cells. A virus that kills its host in this way is called a virulent virus. But there are also viruses that are more temperate, well-mannered guests. Once inside a cell, they hide among the cell's own genes, perhaps producing a few proteins, but otherwise not causing any trouble. Researchers believe that such temperate viruses may occasionally be stimulated to become virulent, or may perhaps be a factor in causing cancer.

Some viruses infect plants; others infect animals. There are also viruses that infect bacteria. These viruses are called *bacteriophages,* or phages for short. (*Bacteriophage* means "bacteria eater.") The phages that infect *E. coli* and other common microorganisms used in the laboratory have been intensively studied, and the complete structures of several of them have already been worked out.

Bacteria and viruses have provided researchers with convenient "living laboratories" from which they have learned a great deal about molecular genetics, much of it applicable to animals and plants as well. The microorganisms are also providing tools for changing heredity, in the rapidly growing but still infant science of genetic engineering.

Losers in the Lottery: the Victims of Genetic Diseases

"Should we have children?" Many couples have anxiously asked themselves that question, worried about the possibility that a serious illness or deformity in some family member might be passed on to their children. Such worries are not new. A few generations ago, parents nervously examined their children for signs of abnormality if Aunt Alma had "weak lungs" or Grandpa was "a bit queer" in his old age. Some people conscientiously decided that it was their duty to remain single to avoid passing on the hereditary "taint." Some states even passed laws requiring sterilization of people with undesirable conditions believed to be hereditary— including people with tuberculosis, epileptics, the mentally retarded, and "criminals."

We know now that many of the "hereditary diseases"

people used to worry about are not hereditary at all. The explosion of new genetic knowledge in recent years, however, has provided insights into more than 2,000 diseases that are now known to be genetic. In many cases medical researchers have been able to discover the manner in which they are inherited—for example, the sex-linked inheritance of the recessive gene for hemophilia. In some cases, new knowledge has enabled scientists to develop tests that can be used to determine whether people wishing to have children are carriers of a defective gene or whether children will suffer from a hereditary disease—even before birth!

Genetic diseases may show their effects at various times of life. The most serious cause death before birth—perhaps before the mother even realizes she is pregnant. Many birth defects are hereditary disorders—mistakes in the chromosomes that cause the development of the child to go wrong some time before birth. They may be relatively minor variations, like an extra finger or toe, which can cause the person to feel "different." (Ann Boleyn, the second wife of King Henry VIII of England, is said to have had six fingers on one hand, and in portraits she is shown hiding a hand in a fold of her skirt or behind some object.) A cleft palate is a more serious birth defect, which must be repaired by surgery if the child is to be able to talk and breathe properly.

Spina bifida is a tragic hereditary birth defect in which the spinal cord does not develop properly and the spine's protective tube of bone does not completely form around it. The opening in the spine can be corrected by surgery, but children with this defect are usually mentally retarded to some degree. They may also suffer from other problems, such as paralysis of the legs or "water on the brain," a filling of the skull with fluid that can crowd and damage the delicate brain tissues. An even more serious birth defect, called anencephaly, is the failure of the brain and skull to develop. Infants with this birth defect die before birth or soon afterward.

Down's syndrome, a rather common birth defect, shows its effects at birth. Down's children are sometimes called mongoloids because of a somewhat "oriental" appearance of the eyes. They also have a broad nose and a protruding tongue. This typical facial appearance goes along with a short stature, an unusual pattern of hand- and fingerprints (there is a prominent crease running all the way across the palm), and mental retardation. Some have other serious abnormalities as well, such as heart defects and intestinal blockage. Many Down's syndrome patients die before they are twenty.

When a genetic disease is diagnosed at birth, at least the parents can adjust to the problem and make plans for treatment that may help the child. But there are a number of serious metabolic diseases—errors in the chemical reactions of the body—that do not show themselves until sometime later. The baby appears normal and healthy for a while, but then things begin to go wrong. He or she may not develop as fast as a baby should; or a child who has been walking and playing may gradually become unable to walk, stand, or even sit up. The parents helplessly watch their child's brightness fade away. Depending on the particular metabolic disease involved, the child may live on to be a mentally retarded adult who can perform only the simplest tasks and needs special care and supervision. Or the child may die at the age of two or three, or later, in the teen years.

One of the most upsetting of the genetic diseases is Huntington's chorea, a degenerative disease for which no cure is yet known. The person seems healthy, then gradually begins to develop disturbing symptoms. Twitching movements of the limbs and personality changes are early signs. As the disease develops over a period of years, the victim develops an almost constant movement, with twisting and stretching. Helplessness and an agonizing death come at the end of the struggle. Huntington's chorea is inherited through a dominant gene, which means that every child of an affected person has a 50 percent chance of inheriting the disease. Unfortunately, the first symptoms usually do not appear until the thirties or forties, when a person is likely to have already married and had children. Popular singer Arlo Guthrie is one of the people now living under the threat of Huntington's chorea; as a child he watched his father, folk singer Woodie Guthrie, die of the disease.

Huntington's chorea is a rather rare disease. But medical researchers believe that there are also hereditary factors in some of the major killers: cancer, heart disease, and diabetes, which together account for most of the deaths of the middle-aged and elderly. Particular types of cancer seem to run in certain families. Some people apparently inherit an inability to utilize the fats in foods properly and are likely to die of heart attacks in their thirties or forties. People who have close relatives

with diabetes have a higher-than-average chance of
developing the disease themselves. But the inheritance
of tendencies for these common diseases is rather
complicated and is not completely understood yet.
(Indeed, doctors now believe that the form of diabetes
that strikes young people is actually the result of viral
infection, although there may be some hereditary influence, too.)

A number of hereditary "mistakes" can result in
genetic diseases. One kind of mistake occurs in the
number of chromosomes. Humans normally have a set
of 46 chromosomes, or 23 pairs. Actually, you have
a set of perfectly matched pairs of chromosomes only
if you are female. For in humans, as in many other
animals, sex is determined by a special pair of
chromosomes, called the sex chromosomes. The female
has two matching sex chromosomes, which are called X
chromosomes. The male has an X chromosome, too—
but only one. His other sex chromosome doesn't
match—it is a smaller one, called a Y chromosome. The
Y chromosome contains genes for the development of
male sex organs and also for the special hormones that
produce other characteristics such as a hairy chest and a
deep voice. The X chromosome contains the genes for
female sex characteristics (these genes are normally
inactive if a Y chromosome is present). Since it is a
fairly large chromosome, it also contains genes for other
traits, such as blood-clotting factors. A mother has only
X chromosomes to contribute to each of her egg cells,
since her sex genotype is XX. A father, who has an XY
genotype, contributes an X chromosome to some sperm
cells and a Y chromosome to others. So when a sperm

and an egg join to start the life of a new human being, the new cell is equally likely to be XX or XY.

Normally, when sperm and eggs are being formed, the pairs of chromosomes separate evenly, and each new germ cell receives only one member of each pair. But sometimes something goes wrong. Instead of separating as the cell divides, a particular chromosome pair may stay together. Both chromosomes will enter one of the new cells that is formed, while the other new cell will not receive any chromosome from that pair. Thus, one germ cell will have 24 chromosomes, while the other will have only 22. If one of these faulty germ cells happens to be involved in the start of a baby, the new person will have a chromosome set of 47 or 45 chromosomes. Both variations can mean trouble.

A failure of the chromosomes to separate properly during cell division is referred to as *nondisjunction.* Some chromosome sets with a missing chromosome or an extra one result in a spontaneous abortion or miscarriage early in pregnancy: A lack or an oversupply of key genes makes the cells so abnormal that usually the embryo cannot develop, and it is expelled from the mother's body long before the normal time of birth. Abnormal chromosome numbers are found in about 40 percent of the miscarriages that occur during the first three months of pregnancy.

Some children with abnormal chromosome numbers do manage to survive to birth or beyond. The most common condition of this kind is Down's syndrome, which is also called trisomy-21 because sufferers have three ("tri") of chromosome number 21, instead of the usual pair. Other serious hereditary diseases are caused

AUTOSOMES: 44 PAIRED ACCORDING TO SIZE

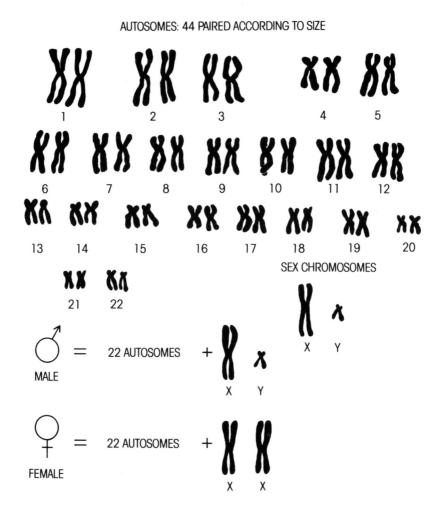

SEX CHROMOSOMES

NONDISJUNCTION DURING CELL DIVISION:

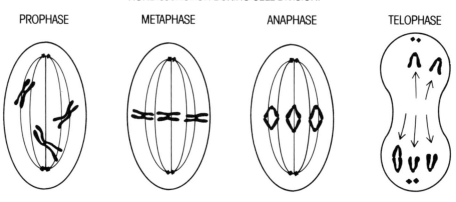

PROPHASE · METAPHASE · ANAPHASE · TELOPHASE

by trisomy of chromosomes 13 and 8. Nondisjunction involving the sex chromosomes can also cause problems. Usually these are not as serious as problems involving other chromosomes, perhaps because the Y chromosome determines only sex characteristics, and when there are two X chromosomes, the genes in one of them are generally "turned off." People with an XO genotype, or monosomy-X (only one X chromosome and no Y chromosome) are females, but they do not mature sexually unless they are given doses of female sex hormones. Even with such treatment, they cannot have children. They also have a typically short stature and a webbed neck; but unlike Down's patients, they are not usually mentally retarded. Trisomy-X—three X chromosomes—usually produces no apparent problems. People with this genotype are normal females.

The XXY genotype produces tall, rather feminine-looking men whose sex organs do not mature enough for them to be able to father children. There has been a great deal of controversy about the XYY genotype. It was widely discussed in the news when doctors reported (incorrectly, as it turned out later) that the notorious murderer Richard Speck had this genotype. Studies have shown that the XYY genotype occurs surprisingly often in inmates of prisons and mental institutions, and researchers speculated that the extra Y chromosome makes a man a "supermale," extra aggressive, a real

The human karyotype is shown above. A male has 22 pairs of autosomes, plus an X chromosome and a Y chromosome. A female has 22 pairs of autosomes, plus a pair of X chromosomes. As shown below, nondisjunction can cause the daughter cells to have too many or too few of a particular type of chromosome.

"criminal type." But then it was discovered that there are also many perfectly normal, law-abiding men who have an XYY chromosome makeup.

Errors of chromosome number are only one type of abnormality that can arise when the germ cells are being formed. While the DNA is replicating (reproducing), a gene may be accidentally duplicated, or a fragment of a chromosome may be lost. Part of one chromosome might break off and become attached to another chromosome, an abnormality called *translocation*. When the cell divides, there are two possible results. The chromosome that lost a fragment and the chromosome to which the extra piece was attached might happen to go to the same germ cell. This is called a balanced translocation, and if that germ cell contributes to the formation of a person, there will be no apparent effect. All the needed genes will be there, in just the right amounts, even though they are not all attached to the proper chromosomes. The other possibility is that when the cell divides, the chromosome that is missing a fragment will go to one of the new cells formed and the chromosome with an extra piece will go to the other. Depending on which particular cell contributes to the formation of a new person, there might be a sort of trisomy (too many genes corresponding to the extra piece) or a monosomy (only a single dose of the genes corresponding to the missing fragment). Even a balanced translocation can cause trouble when the person has children; for when his or her germ cells divide, they in turn might produce sperm or eggs with an unbalanced chromosome set.

Scientists have estimated that our 46 chromosomes

contain a total of about 100,000 genes. Each of these genes carries the code for a particular protein—perhaps a hormone or an enzyme that helps to control the body and its reactions; perhaps one of the body's building materials, such as the protein in hair or nails. The set of genes that we carry has been selected over millions of years of trial and error. The strongest, healthiest, and most able human beings have tended to have more children and pass on their genes. So it seems logical that a change in a gene is likely to make it less, rather than more, effective.

It doesn't take much of a change to make a difference in a gene's effectiveness. A break in the DNA chain might leave the gene unable to transcribe its message into RNA. The addition of a small fragment, the loss of a few nucleotides, or even a chemical change in just one of the nitrogen bases would alter the message coded in the gene. Some mutations of this sort can reduce the message to nonsense. Imagine, for a moment, what would happen if the last sentence lost its first letter, and the spacing of all the words was shifted over by one letter. It would become: Omem utationso ft hiss ortc anr educet hem essaget on onsense.

A substitution of a single nitrogen base might have no effect at all: The genetic code, with more than one codon corresponding to the same amino acid in most cases, has a built-in compensation for a certain amount of error. Even if a substitution of bases does result in a different amino acid's being built into the protein, the change might occur in a part of the molecule that does not make much difference. Indeed, there is a good deal of variation in the "normal" proteins of different individuals. But the change might also happen to fall in a crucial spot. Sickle-cell anemia is a hereditary disease in which the red blood cells have a tendency to collapse from their normal doughnutlike shape into a sickle shape, painfully clogging the blood vessels and failing to carry enough life-giving oxygen through the body. This condition has been traced to a change in a single amino acid—only one out of about 300—in the hemoglobin molecule.

For many genetic diseases, the manner of their inheritance has been worked out, and doctors can predict the risk of passing on the condition. Non-disjunction, for example, tends to occur most often in older women, during the final division of the egg cell. In women under thirty, there is a likelihood of only one chance in 750 of having a child with an abnormal chromosome number. But later—especially after thirty-five—the odds rise sharply. In the 40–44 age group, the chances of having a baby with an abnormal chromosome number are up to one in 50, and over age forty-five they are one in 20. (In each age group, about half the abnormal cases will be children with Down's syndrome.) In addition to the mother's age, heredity may also be a factor, since some families have a much higher than expected number of trisomies, and a woman who has already given birth to a Down's child has a higher-than-normal risk of having more. Perhaps in such women the genes that normally control the correct separation of chromosomes do not work effectively. In some cases the condition is due to a chromosome translocation.

Some genetic diseases, caused by mutation of a gene,

are inherited as dominant traits. This class includes achondroplastic dwarfism (dwarfs with a relatively normal-sized head and trunk but very short arms and legs), Huntington's chorea, the "India Rubber Man" syndrome, porphyria (periods of insanity of King George III, who ruled in England when the American colonies won their independence, were due to this disease), and retinoblastoma (a form of cancer that affects the eyes). Children of parents with a dominant genetic defect have a 50 percent chance of receiving the chromosome with the bad gene and thus inheriting the disease. This probability is the same for each child, regardless of whether earlier children in the family were affected or not. If that does not seem reasonable to you, try flipping a coin. At each flip there will be the same probability that the coin will come up "heads," no matter how many heads you have flipped before. Just as you might, by chance, turn over three heads in a row, so a parent carrying a dominant mutation might have three affected children in a row—or none at all. And the next time, there will still be the same 50 percent probability.

Recessive mutations can also cause genetic diseases. But their effects will not appear unless the person inherits the same recessive gene from both parents. If only one recessive gene is present, the normal gene on the corresponding chromosome will turn out enough of the normal product to compensate. Heterozygous carriers of the sickle-cell trait, for example, produce about 45 percent of the abnormal hemoglobin; but since the rest of their hemoglobin is normal, their red cells do not "sickle." Although recessive mutations occur just as often as dominant ones, the diseases they cause are rarer, and the probability of inheriting them is much lower. Marriage of close relatives, such as first cousins, increases the odds of combining recessive mutations in the children, since the parents share a number of genes in common. Metabolic diseases such as phenylketonuria and Tay-Sachs disease are inherited through recessive genes.

Sex-linked recessive conditions are a special case. Here the mutated gene is found on the X chromosomes. Girls have two X chromosomes, so they will show the symptoms of the disease only if they inherit the mutated gene from both parents, just as with other recessive conditions. But for boys the situation is different. A boy receives an X chromosome from his mother and a Y chromosome from his father. If his mother was carrying the recessive gene for a condition such as hemophilia or Duchenne muscular dystrophy on one of her X chromosomes, the boy has a 50 percent chance of receiving the chromosome with the faulty gene. If he does, he will have the disease, even though he has only one gene for it—for his Y chromosome does not contain a normal gene to compensate for it. A man with an X-linked disease can pass on the faulty gene only to his daughters, who will be carriers without outward symptoms (unless their mother was also carrying the gene). His sons cannot inherit it.

The knowledge that researchers have gained about genetic diseases and how they are inherited is now being put to practical use. A new type of medical specialist, the genetic counselor, is giving couples the facts they need to know in deciding whether they should have children. In many cases, teams of genetic counselors can provide concrete answers to the question "Will our baby be all right?"

The experience of a friend of ours was typical. Several years ago, when her first daughter was five, our friend gave birth to a child with Down's syndrome. The first few months after Erika's birth were an emotionally exhausting time for the parents—first doubting the diagnosis ("Are you *sure*? She doesn't *look* that abnormal."), then gradually adjusting to the fact that their new daughter would never be able to do all the things that normal children could, probably would often be ill and need special care, and might not survive to adulthood. They had to face the agonizing decision of whether to care for Erika themselves or place her in an institution. (They decided to try keeping her with them for a while and "see how it goes.") When they were finally able to spare some attention from the problems of Erika's future to thoughts of their own, they realized they were facing an important question: Did they dare to have another child? They had deeply longed for a son, and still did. But were they prepared to take the risk of having another child like Erika? Just what was the risk?

Their doctor referred them to a genetic counseling team. A complete family history was taken: Both our friend and her husband were questioned closely about whether any relatives in the past few generations had had Down's syndrome. Blood samples were taken, not only from the parents, but also from Erika and her older sister. Then the family waited for the results of the laboratory tests. White blood cells from each sample were grown in laboratory cultures. The cells multiplied for three days in the special tissue culture medium, and then cell division was stopped with a drug called colchicine. Cells were spread out on microscope slides, stained to show the chromosomes, and then examined and photographed under a microscope. A karyotype of each cell was prepared by carefully cutting each chromosome out of a blown-up photograph and sorting all the chromosomes of the set according to size and shape.

The results of the test brought encouraging news. Erika's cells showed the typical trisomy-21, but there was no sign of this abnormality in any of the cells from our friend, her husband, or their older daughter. Nor were there any translocations in any of the karyotypes. Since our friend was in her late twenties—not in the high-risk group—and Down's syndrome had not appeared in any of her relatives, the genetic counselors advised that there was probably no special risk that the couple would have another Down's child. Erika's birth defect was apparently a one-in-a-thousand chance, and the genetic dice would roll again at the same odds for the next child.

Our friends decided to try again. But partway through the pregnancy they began to worry. They had lost in the "genetic lottery" once; what if it had happened again? How could they face the months of uncertainty that lay ahead?

The doctor suggested *amniocentesis,* a procedure that has been developed into a valuable tool for genetic counseling. Before birth, the baby-to-be grows inside its mother's abdomen, suspended in a fluid-filled bag that protects it from bumps and shocks. This bag of tissue is called the amnion, and the fluid that surrounds the growing fetus is called amniotic fluid. It contains water, dissolved salts and other chemicals, and also various waste products and sloughed-off cells from the fetal body. In amniocentesis, a long, hollow needle is carefully inserted through the mother's abdomen into the amniotic cavity, and a small sample of amniotic fluid is drawn out. This fluid usually contains some cells from the fetus, and these can be grown in tissue culture and then studied. Chromosome abnormalities can be picked out by staining and examining the cells. The culture can also be used to test for abnormal or lacking enzymes that might indicate the presence of a metabolic disease.

In our friend's case, the laboratory tests were reassuring. The culture showed no signs of trisomy-21 or any other chromosome abnormalities. Their new child would not be a Down's child. And as an extra benefit, they found out several months in advance that the child they were expecting was a boy. After that, it seemed like no time at all until they were beaming down at their normal, healthy, newborn son.

Genetic counseling and amniocentesis are usually used for people in "high-risk" categories—couples who have already had an abnormal child, or know of a genetic disease that runs in their family; or mothers in their late thirties or forties, who have a much greater than average chance of having children with chromosome abnormalities. In the great majority of cases, the results of amniocentesis bring great relief to the parents by revealing that their child will be normal. Even for a woman over forty-five, for example, with a high risk of one in twenty that her child will have an abnormal chromosome number, there are nineteen chances in twenty that it will not.

When tests before birth or shortly afterward do reveal a genetic disease, in some cases prompt treatment can minimize the symptoms and help the child to develop relatively normally. For example, researchers studying one metabolic disease in the early 1970s made an interesting discovery. The disease, which caused retarded growth and mental development, was thought to be due to the lack of a key enzyme. But tests showed that children with the disease had plenty of the enzyme in their blood and tissues. It turned out that they were lacking something else: an ability to use vitamin B_{12} properly; for this vitamin is known to work with the enzyme in the body. The research team soon had a chance to put their new knowledge to practical use. Amniocentesis showed that a woman was carrying a child with the disease. If they waited until the child was born, it would be doomed to be hopelessly retarded. So they decided to treat the fetus right away, even though it had been developing inside its mother's body for only 19 weeks. (A normal pregnancy lasts 39 weeks.) They gave the mother large doses of vitamin B_{12} since nourishment passes from the mother's blood into that of her growing baby through the placenta. Enough of the extra

In amniocentesis, cells shed from the fetus into the amniotic fluid are grown in culture and analyzed for chromosomal and other abnormalities.

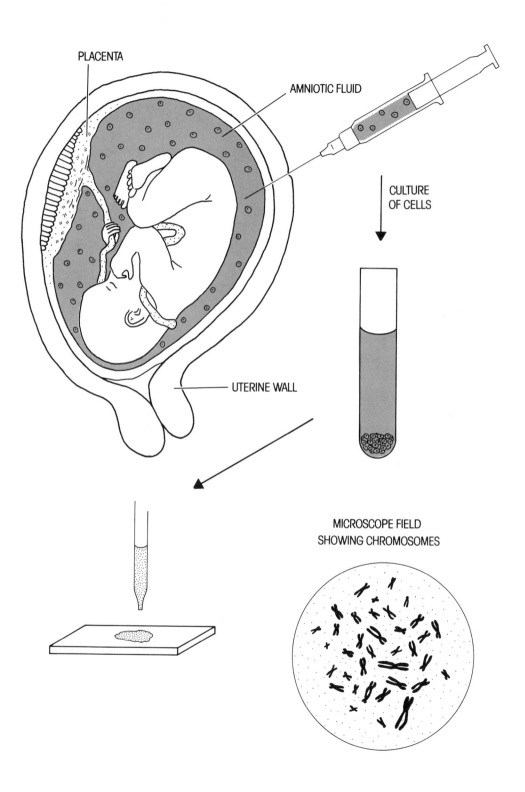

PLACENTA

AMNIOTIC FLUID

CULTURE
OF CELLS

UTERINE WALL

MICROSCOPE FIELD
SHOWING CHROMOSOMES

vitamin B_{12} passed into this baby's blood to make its enzymes work effectively. The baby was normal at birth and, with a special vitamin supplement, continued to grow and develop normally afterward.

Amniocentesis would not be suitable for large-scale testing of the general population, for it must be performed by a skilled surgeon and carries a small risk of injuring the fetus or even bringing on a miscarriage. But some safe, simple, and fairly inexpensive screening tests have been developed for identifying carriers of defective genes and babies who are endangered by a genetic disease. After the birth of our youngest child, for example, the nurse at the hospital gave us a small square of treated paper to place in our baby's diaper after a certain number of days and then send in to the health department in a postpaid envelope. No one had any particular reason to suspect there was anything wrong with our child, and there wasn't; the test was just a routine procedure in one of the largest screening programs ever started. Our state, like most of the others in the United States, had passed a law requiring that all newborn babies be tested for a genetic disease called phenylketonuria, or PKU.

PKU is one of the metabolic diseases. The faulty gene that causes it is recessive, so a person must be homozygous—with a double dose of the gene—to suffer from the disease. About one child in each ten thousand born in the United States suffers from PKU, but about one person in fifty is carrying the gene in heterozygous form. PKU sufferers lack an important enzyme called phenylalanine hydroxylase, which works on an amino acid (phenylalanine) that is found in most protein.

Phenylalanine is a key substance in the body's metabolism. It can be chemically transformed into compounds needed for the work of the brain. Other products of phenylalanine reactions form the dark pigment melanin, which gives color to hair, eyes, and skin. In PKU these reactions are blocked by the lack of the enzyme that the mutated gene would normally produce. (In carriers, the normal gene on the other chromosome of the pair produces enough phenylalanine hydroxylase to compensate; so there are no symptoms.)

What happens when these reactions are blocked? The brain, starved of the chemicals it needs, does not develop properly, and mental retardation and frequent epileptic seizures result. PKU children are typically very fair-skinned and fair-haired, because of the lack of melanin. Meanwhile, the phenylalanine from foods builds up to high levels in the blood and tissues. Some of it is converted to abnormal products called phenyl ketones, which also accumulate in the body. Some of the excess phenylalanine and phenyl ketones spill over into the urine. (That's the basis for the name of the disease—*phenyl-keton-uria*.) These substances give the urine of PKU patients a sort of musty, "mousy" odor that a doctor can recognize immediately. In fact, the disease was first diagnosed in 1934 in Norway, when a mother took her two retarded children to the doctor because they always smelled funny, no matter how hard she tried to keep them clean. The odor came from the urine on their diapers, and chemical tests soon showed that it contained phenyl ketones. With that clue, studies were made of retarded patients in mental institutions, and it was discovered that about one in every hundred inmates of such institutions is a PKU sufferer.

As the biochemical nature of the disease was gradually worked out, it was found that the condition of patients in institutions improved when they were fed a diet from which phenylalanine had been removed: their behavior became more normal, and they had seizures less often. But they were still mentally retarded. Doctors wondered: Is it possible to prevent the mental retardation by feeding PKU children the special diet *before* their brains become damaged? To do that, they had to find a way to identify PKU babies as soon after birth as possible.

At birth a PKU baby seems normal. During pregnancy, while the baby still depends on its mother's body for nourishment, her phenylalanine hydroxylase is able to pass through the placenta and keep the baby's phenylalanine level fairly normal. But after birth a baby's metabolism is on its own. Quickly the phenylalanine level in the PKU baby's blood rises, as it begins to receive nourishment from protein-rich milk. Eventually there is so much of the amino acid in the blood that it spills over into the urine; but there may not be enough for a urine test to be positive until a month or so has passed. Now the screening of infants for PKU is usually done by testing a drop of blood taken from the baby's heel.

In the excitement that followed the development of reliable tests for PKU screening, state after state rushed to pass laws requiring testing of all newborn infants. (About 90 percent of the babies born in the United States are now routinely tested for the disease.) The treatment of PKU babies by feeding them a special diet low in phenylalanine was successful, just as the doctors had hoped. The children on the diet remained normal and healthy, with no signs of mental retardation.

After a while, however, some problems began to emerge. Patients found the PKU diet very "blah"-tasting and boring. When they were very young, they ate what their parents told them to. But then when they went to school and saw all their friends eating forbidden treats like ice cream, hamburgers, and pizza, they began to rebel. Gradually parents and doctors began a general practice of taking the children off the special diet when they were about six, hoping that by that time their intelligence had developed sufficiently. At first it seemed that this was true; but recently there has been some evidence that after years on a normal diet, the high phenylalanine levels cause some loss of intelligence in PKU patients. Their mental abilities do not fall to retarded levels, but in some cases there definitely seems to be a loss. As the first treated PKU patients grew up and began to have children of their own, another problem also appeared: High phenylalanine levels in the PKU mother's body can damage the brain of her developing child. Now doctors advise a woman who has been treated for PKU to start the diet again before becoming pregnant and to keep her phenylalanine intake low during pregnancy.

Despite the problems, PKU screening definitely seems to have been a successful experiment. It has been estimated that the entire PKU screening program in the state of New York costs about $250,000 a year—which is considerably less than the cost of lifetime care for just a few mentally retarded victims of the disease in an institution. In dollars and cents, that seems like a real bargain—and who can put a value on the saving of so many minds that would otherwise have been wasted?

Another mass screening program that has produced some success involves a genetic disease called Tay-Sachs disease (TSD). Named after the doctors who first described and studied it, this condition affects mainly Jews whose ancestors came from Eastern Europe. (Apparently the mutation first arose there and spread through the population.) Among this group, about one person in twenty-five is a carrier, and the disease occurs in one out of about 3,000 infants. (In the general population of the United States, TSD is a much rarer disease, occuring in only about one of every 360,000 births.)

A Tay-Sachs child seems perfectly healthy during the first half-year of its life, developing normally in every way. But then the development seems to go backward. The child stops playing and crawling and babbling; after a while it is too weak and listless even to hold up its head. Gradually the child becomes blind and helpless. As the disease progresses, its body becomes rigid, and loud noises may send it into convulsions and fits of crying. Death puts an end to the child's suffering by the age of three or four. The parents are left to try to pick up their lives again after an experience that has drained them emotionally and cost them as much as $50,000 a year in caring for their afflicted child.

Tay-Sachs disease is a metabolic disease, caused by a recessive mutation of the gene that produces an enzyme called hexosaminidase A, or Hex A. This is an enzyme of the body's fat metabolism. Without it, a fat that is normally found in the brain cannot be broken down and converted to other substances. Instead, it builds up in the brain tissue, clogging and damaging it. The amount of Hex A circulating in the blood is an indicator of how many working genes a person has for this enzyme. A homozygous Tay-Sachs patient has no circulating Hex A at all. A carrier, with one faulty gene and one normal gene, produces about half the normal amount of Hex A.

In the early 1970s, a mass screening program was set up on a trial basis in the area of Baltimore, Maryland. After intensive publicity to inform Jewish people of the area about the disease and its danger to children, centers were set up to take blood samples for testing. Over 60,000 couples were tested, and a number of carriers of the Tay-Sachs gene were identified. They were called in for genetic counseling. It was explained that if only one parent is a carrier, there is no danger that any of the children will suffer from the disease (though there is a 50 percent chance that each child, in turn, will be a carrier of the gene). If both parents are carriers, there is a one-in-four chance of having a child who is doomed to death. For such couples who decided to take the risk and have a child, amniocentesis could let them know in advance whether their child would have the disease, for the fetal cells obtained in the sample of amniotic fluid produce Hex A if the working gene is present. Imagine a couple's relief at hearing that their child is one of the lucky three out of four. Unfortunately, medical science has not yet found a treatment that can help the children who do suffer from Tay-Sachs disease.

Screening programs for another disease have proved less successful. Sickle-cell anemia (SCD) is found

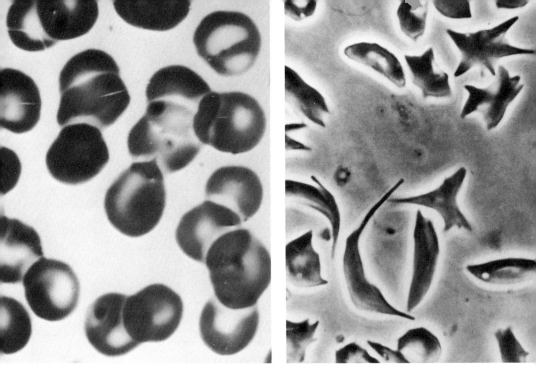

Blood smears, as seen under a microscope. Normal red blood cells are shown on the left and sickle cells on the right.

mainly in blacks. It is believed that the gene became common in Africa because carriers of the trait have some natural protection against malaria, the major killer in the hot regions of the world. Heterozygous carriers of the sickle-cell trait have one gene that produces the normal form of hemoglobin (the protein that gives blood its red color) and one gene that produces a variation called hemoglobin S. Apparently the malaria parasite cannot live comfortably in blood with that mixture. In the United States malaria is not an important health problem. So Americans of African descent carrying sickle-cell genes have lost their advantage and are left with the genetic risk: When two carriers marry, some of their children may have a homozygous sickle-cell genotype. Then only hemoglobin S will be formed, and the red blood cells will have a tendency to change from their normal "doughnut" shape to a sickle shape. Sickle cells tend to clump together, clogging the blood vessels and preventing red cells from getting through. Tissues "downstream" are starved of the oxygen they need, producing painful "sickle crises" that may end in death. (In heterozygous carriers of the sickle-cell trait, there is enough normal hemoglobin to protect the blood from sickling, except under very unusual circumstances, such as high up in an unpressurized airplane.) Among American blacks, about one in twelve is a sickle-cell carrier, and sickle-cell anemia occurs in about one out of each 600 babies born.

When tests were devised to identify hemoglobin S in blood samples, the situation seemed made to order for mass screening and genetic counseling. In some cases such screening programs were set up, but without enough education of the community on the nature of the disease and what the tests meant. Some well-meaning amateurs went around the neighborhoods with blood-testing kits and left many people with the knowledge that they were "carriers" but no understanding of what that meant or what to do about it. Members of the black community worried that the diagnosis of sickle-cell carrier might be entered in data banks and used as a basis for denying jobs or insurance coverage. And growing feelings of black self-awareness and pride led some leaders of the community to react angrily to suggestions that some black couples should not have children.

Sickle-cell anemia is not one of the several dozen diseases that can now be diagnosed by amniocentesis, because hemoglobin is produced only by red blood cells, and these cells are not shed into the amniotic fluid. If two carriers do decide to take the chance and have a child, there is no way to determine until after the baby is born whether it actually has the disease. When a child with SCD is diagnosed, there is little that can be done: Though several promising treatments are now being tried out, there is still no sure cure.

How *do* you cure a genetic disease? Sometimes surgery provides an answer. Surgeons can skillfully correct harelips, cleft palates, and other abnormalities of development. In retinoblastoma, the surgeon may have to remove one or both of the child's cancerous eyes. It seems a horrible treatment, but it can save the child's life.

Studies of the body's chemistry have suggested fruitful approaches to treating many genetic diseases. Treatments may supply a missing hormone or other biochemical, or prevent the buildup of some harmful product. Injections of sex hormones can help to normalize people with abnormal numbers of sex chromosomes. The injections supply the chemicals that lacking genes would have produced. Regular injections of insulin permit millions of diabetics to live a relatively normal life, even though their pancreas does not produce enough insulin or their body cells cannot effectively use the insulin they do have. Sufferers from hemophilia, the "bleeder's disease," can be helped by injections of Factor VIII, a natural clotting agent that is collected from donated blood.

Treating PKU with a special diet that prevents the buildup of phenylalanine is an example of the second kind of biochemical approach. Several other metabolic diseases can also be treated with special diets. In Wilson's disease, a rare recessive condition, a normal-seeming child or teen-ager suddenly begins to lose muscle control and coordination. Later, cirrhosis of the liver develops, and if the disease is not treated, the victim soon dies. Researchers have not yet found the particular enzyme involved, but they have discovered that the problem is an inability to use copper properly. This metal, which is important in a number of chemical reactions in the body, accumulates in such large amounts that it becomes a poison. A drug called D-penicillamine has the ability to bind tightly to copper and help to flush it out of the body. Nowadays, if Wilson's disease is diagnosed early enough, before there has been any permanent damage, penicillamine brings dramatic improvement. Many Wilson's disease patients are now leading normal lives with the help of this drug, and some have even had trouble-free pregnancies and borne healthy children. But they must continue to take the drug all their lives.

All these treatments have been developed on the basis of knowledge of how the normal body works and how a faulty gene can set the cells' biochemistry awry. But supplying lacking chemicals or introducing drugs to try to compensate for reactions that have gone wrong seems like a rather artificial approach. Medical researchers have dreamed about actually zeroing in on the faulty genes and correcting them, so that the body will make its own cure. Advances in the infant science of genetic engineering are promising that some of these dreams may be translated into reality.

Reading the
Blueprints of Life:
the Keys to
the Kingdom

Biomedical researchers are now thinking seriously about prospects for re-engineering our heredity—rewriting the blueprints in the cells to correct errors and perhaps even to improve on nature's designs. If these are not to be just idle pipe dreams, researchers must be able to do two things: to read the blueprints of life in their chemical code and to use that code to make changes in the existing information molecules or even to build new ones to order.

When you think that just a few decades ago scientists did not even know what kind of chemicals made up the molecules of heredity, we have already come an amazingly long way. In 1976 molecular biologists announced that they had figured out the complete chemical structure of an organism: MS2, a virus that invades the bacterium *E. coli.* MS2 is not a very complicated organism, to be sure. It consists of an RNA core with only three genes and the three proteins coded by the genes. Some scientists would debate whether MS2 is really a living organism. Even if it is, the achievement may not seem very impressive when we compare this tiny creature with a simple bacterium like *E. coli* with about a thousand genes, a fruit fly with about 5,000, and a human with a set of perhaps 100,000 different genes in each cell. Yet MS2 was a first step, and several more complete virus structures (including forms with DNA) quickly followed. In the years ahead, we will see a flood of announcements of complete chemical structures of ever more complex organisms. Even the human genome—our whole set of 46 chromosomes—is now rapidly revealing its secrets to the chemical detectives.

The science of reading the chemicals of life has grown rapidly from its beginnings only a few decades ago. The pioneer who blazed the trail was a British biochemist named Frederick Sanger. Working for ten years with carefully selected chemical reactions, new techniques for separating chemicals, and some inspired guesswork, Sanger figured out the complete sequence of all 51 amino acids in the insulin molecule. The announcement of his achievement in 1954 took scientists by surprise. Excitement spread throughout the world. Many researchers had thought the complete amino acid sequence of a protein would *never* be worked out—that proteins were just too large and complicated to be knowable.

Of course, insulin is not a very large protein molecule. But once its structure had been worked out, the same techniques could be applied to more complicated proteins. In the quarter century or so that has passed since Sanger's announcement, literally thousands of protein structures have been determined, including some of giant proteins more than a hundred times larger than insulin. New laboratory techniques and the growing use of computers have made it possible for researchers to do in a matter of weeks, or even days, the kind of job that took Sanger ten years. Indeed, there are even automatic sequenators that can take a protein molecule and cut off and identify the amino acids, one by one.

Frederick Sanger won a well-deserved Nobel Prize for his work on the structure of insulin. After that he left it for others to work out the details of protein structures and turned his questing mind to another area: the structure of nucleic acids. DNA and RNA molecules are even larger and more complicated than proteins. Sanger devised new techniques for cutting up the long chains of nucleotides and analyzing the pieces, in order to build up a picture of the nucleic acid structure. Using Sanger's techniques and others invented by Walter Gilbert and his associates at Harvard University, molecular biologists are now making extraordinary progress in reading the blueprints of life. In 1976 Gilbert noted that, just three years before, he and his research team had taken two whole years to determine the nucleotide sequence of a piece of DNA twenty nucleotides long. By 1976 they were able to determine such a sequence in a day. Now the work is going even faster.

Meanwhile, another revolution is going on in the area of gene synthesis. Researchers are learning to duplicate in test tubes and culture dishes the building of nucleic acids that goes on in living cells. They are even gaining the skill to put together genes that never existed before—to write their own instructions into the nucleic acid plans.

Late in 1967 an exciting announcement was made. Molecular biologist Arthur Kornberg had taken DNA isolated from a small virus called ϕX174, added enzymes and an assortment of nucleotides, and produced a ring of DNA that was a faithful copy of the original virus DNA. President Johnson hailed Kornberg's work as "an awesome accomplishment," and some enthusiastic scientists described it as "essentially the synthesis of life." That description was a bit exaggerated— Kornberg did not really synthesize life; he only took the gene-making components from living cells and provided suitable conditions for the reactions to take place in a test tube.

Just a few years later, in 1970, another molecular biologist announced that he had actually synthesized a gene—"from scratch"—without using a natural nucleic acid as the pattern. Nobel Prize winner Har Gobind Khorana, at the University of Wisconsin, put strings of nucleotides together in a precisely determined sequence to form an artificial gene 77 nucleotides long. He used an ingenious technique that he had developed, called the "sticky end" method. He started by chemically bonding nucleotides together into two short single strands. They corresponded to a small portion of the chemical sequence of a yeast gene whose structure researchers had already determined. One of the short nucleotide chains corresponded to one strand of the yeast DNA and the other short chain to the other. The sequences Khorana synthesized were carefully selected from the known nucleotide sequence of the yeast gene so that they overlapped partly, but not completely. When he mixed the two fragments together, they lined up neatly, forming hydrogen bonds in the proper pairing—A with T and C with G. The result was a short piece of DNA, which was double-stranded in the middle but had a

Khorana used a "sticky end" technique to synthesize a gene, adding matching single-stranded fragments one at a time to build up the DNA molecule. "Sticky ends" also play a key role in recombinant DNA techniques.

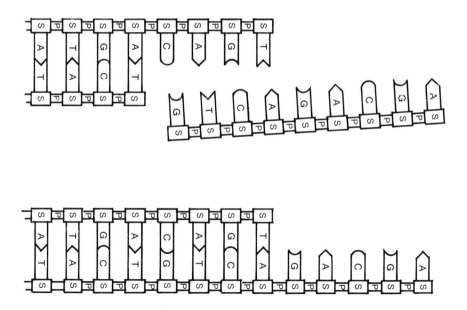

small single-stranded portion dangling from each end. These were the "sticky ends." Next Khorana built up another small chain of nucleotides, this time corresponding to the next part of the sequence on one of the yeast DNA strands. When he added this new piece, the nucleotides at its end paired up smoothly with those on a "sticky end" of the DNA he was building. An enzyme called DNA-joining enzyme was used to bond the two fragments firmly together. (In living cells this enzyme works to repair breaks in DNA.) Khorana continued building up his gene, piece by piece, first working on one strand and then on the other in turn, until the gene was complete.

Khorana's "sticky end" technique quickly became a standard method for making DNA molecules. Many genes have now been synthesized in the laboratory. Khorana himself achieved another "first" in the field by synthesizing a working *E. coli* gene, complete with the appropriate "on-switch" and "off-switch" that permit it to be transcribed into an active transfer RNA. So far researchers have been concentrating on reproducing real genes from living organisms. But Khorana's method could also be used to create new genes with structures that have never occured in nature.

Just about the same time that Khorana announced his first gene synthesis, a revolution occurred in biological theory which opened up a whole new approach to the synthesis of genes. By the late 1960s, molecular biologists had grown rather comfortable in the knowledge that they had firmly worked out the chemical basis of heredity and the manner in which the instructions coded in DNA are translated into the variety of chemicals and reactions in the cell. Everyone knew that DNA forms RNA, RNA forms protein, and the proteins do everything else. This was considered the proven, time-tested "central dogma" from which all the other theories of molecular biology flowed. But two researchers in different laboratories, Howard Temin of the University of Wisconsin and David Baltimore of the Massachusetts Institute of Technology, were thinking subversive thoughts. They were studying viruses that have been shown to cause cancer in animals. These viruses, known as C-type viruses, contain no DNA, only a core of RNA. Yet when they invade a cell their RNA "genes" can direct the cell's synthetic systems to build new virus particles. Temin and Baltimore discovered that the C-type viruses contain a special enzyme called *reverse transcriptase.* When the virus invades a cell, this enzyme directs the cell to build a gene of DNA, using the RNA of the virus as a template. In this reverse transcription, a DNA copy of RNA is made, instead of the other way around. The new DNA is then inserted into the host cell's chromosomes and works as though it were a normal gene.

The discovery of reverse transcriptase provided an important insight into the problem of cancer. It also shook the foundations of molecular biology. Temin and Baltimore received the 1975 Nobel Prize in physiology and medicine for their work. Meanwhile, after their fellow researchers recovered from their shock and realized they would have to be less dogmatic about their "central dogma," they discovered that reverse transcriptase provided a handy shortcut to making genes. A team at Harvard University borrowed some of the enzyme and set to work building DNA from the messenger RNA for a rabbit hemoglobin. In the laboratory the researchers did the same thing that a C-type virus does inside a living cell: They made a DNA copy of RNA. There were problems along the way. The DNA they first obtained was a copy of the RNA, all right, but it had only one strand, like a railroad track with only one rail. It also had an odd little half-loop on one end. After nine more months of work, the Harvard team solved both problems. They found another enzyme, called DNA polymerase I, which used the extra loop as a starting point for building the second strand of DNA; and then still a third enzyme, to trim off the loop from the finished product.

The Harvard researchers had developed another powerful tool for making genes. Now a number of messenger RNAs can be isolated in fairly large quantities. When cells are actively making a particular protein, many molecules of messenger RNA carrying the code for that protein are produced in the nucleus and sent out to the ribosomes to serve as patterns. With plenty of a particular RNA molecule to work with, researchers can use reverse transcriptase and DNA polymerase to reconstruct a gene corresponding to it, even if its nucleotide sequence is not yet known.

While molecular biologists have been learning to read the chemical blueprints of life and to make their own genes in the laboratory, other research teams have been making good progress in mapping chromosomes and figuring out what particular genes do. One powerful technique that they are using is called *cell fusion*.

Most of the body's cells normally have only one

nucleus, which contains the control centers that direct the cell's activities. But for a long time—dating back to 1895—there have been occasional reports that cells in diseased tissues sometimes have more than one nucleus. They look as though two or more cells had joined together. It was not until the 1960s that geneticists discovered that this is what really does happen, and developed ways of making cells fuse.

Henry Harris, a pathologist at Oxford University in England, was a pioneer in this field. He discovered that viruses can cause two cells to fuse and mingle their contents. After a long process of testing various viruses, he finally selected a strain called the Sendai virus. Treating this virus with ultraviolet light destroys its ability to infect cells, but it can still make them fuse. When two cells in contact are treated with Sendai virus, their outer membranes open and they flow together, forming a single large cell. This new cell contains all the contents from the two original cells, including a nucleus from each of them. First Harris used his technique on ordinary tissue samples. But then he tried some exotic variations. He tried mixing human and mouse cells, then treating the mixture with Sendai virus. Some of the mouse cells fused with mouse cells. Some of the human

Human cells can be fused with cells of other species to form hybrid cells. When human-mouse hybrid cells are grown in culture, over the generations chromosomes (especially the human chromosomes) are gradually lost. [Some of the chromosomes from each set have been omitted in the drawing, for greater clarity.]

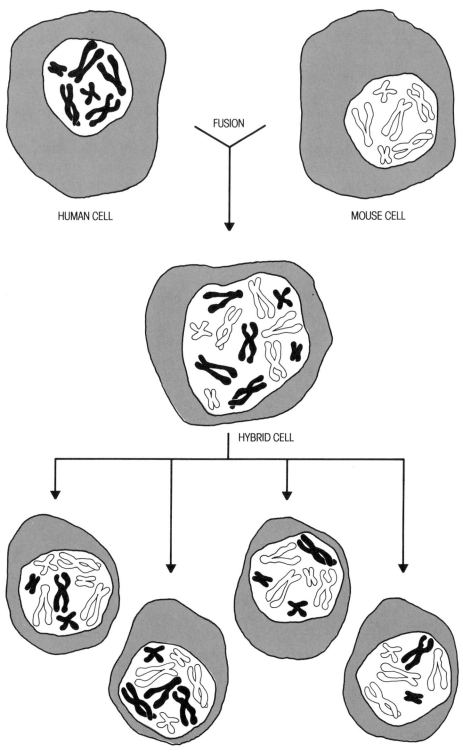

FUSION

HUMAN CELL

MOUSE CELL

HYBRID CELL

HYBRID CELLS AFTER GENERATIONS OF GROWTH IN CULTURES

cells fused with human cells. But there were some cases where human cells fused with mouse cells, forming a mouse-human hybrid! Harris also produced a number of other hybrid cell variations, including mouse-rat, mouse-hamster, and mouse-chicken combinations.

Culturing the hybrid cells produced some interesting and valuable results. The hybrid cells grow and divide, much like other cells in culture. But with each division, chromosomes are lost. In the mouse-human hybrids, human chromosomes are usually the first to go. There does not seem to be any standard pattern to this loss. One cell might lose the human chromosome 22, for example, while another cell from the same original hybrid cell parent might lose the human chromosome 3. The loss of chromosomes goes rapidly at first, but after a number of cell divisions it slows down. Eventually fairly stable cell lines are formed, containing mainly mouse chromosomes and a small number of human chromosomes. Researchers know that both mouse and human chromosomes are working, for chemical tests of the cells show that both mouse and human proteins are present.

Hybrid cells produced by cell fusion are a wonderful tool for mapping chromosomes. Researchers can test cells of various hybrid lines for a particular protein—perhaps a typical human enzyme. If they find it, they examine cells of that line under a microscope and identify the chromosomes present. Since only a few human chromosomes remain, they know that the gene for the human protein they are studying is located on one of them. Other hybrid cells that produce the protein may have a different assortment of chromosomes, and by comparing the results, some possibilities can be eliminated. Gradually the choice is narrowed down to just one possible chromosome for a particular protein. Then, by using cells with unusual chromosomes—ones that have lost a portion or transferred a segment to another chromosome—the researchers can narrow the search even further, locating the gene on a particular part of the chromosome. X rays and chemical mutagens can be used to produce chromosome damages, providing more information for the gene mappers.

In 1975 Frank Riddle of Yale University's School of Medicine reported that more than a hundred human genes had already been mapped, mainly using the cell fusion technique. He was amazed at the explosive progress in his field: The year before, the total number of human genes mapped stood at only fifty, and back in 1970 only one or two had been located. Since 1975 progress has continued, and hundreds of human genes have now been mapped.

Cell fusion is finding a number of other uses as well. Researchers are using hybrid cells to study the functions of tissues and to learn how genes are turned on and off at various stages of development. Each cell in the body contains a complete set of genes, but only the genes for its particular jobs can work. In skin cells, for example, there are genes for making muscle proteins, liver enzymes, and the hormones of the pancreas; but all these genes seem to be permanently "turned off." Yet when mouse liver cells are fused with human skin cells, the hybrids produce some *human* liver proteins. Apparently the fusion somehow turns on the liver-cell genes that were included in the skin cells' chromosome set. Frank

Riddle suggests that development of this technique may make it possible to test for many more genetic diseases through amniocentesis. Until now, for example, doctors have been unable to test for sickle-cell anemia before birth, because hemoglobin S is produced only by blood cells, not by the skin cells shed by the fetus into the amniotic fluid. However, if these skin cells could be fused with mouse blood cells and cultured, they might produce some blood proteins that could be tested.

Researchers are also excited about the possibility of using cell fusion for real genetic engineering—to correct genetic defects. Samples of the person's own cells could be fused with cells containing genes that produce the lacking enzyme. Then the hybrid cells could be introduced into the person's body to multiply and supply the lacking chemical. Nelson Levy's research team at Duke University Medical Center is working on experiments along this line, using mice lacking the gene for a key part of the immunological reaction that protects the body from disease germs.

An unusual variation of cell fusion is providing more information on how the genes work. In the early 1970s it was reported that when animal cells are treated with the drug cytochalasin B, a curious thing happens: The nucleus of the cell pulls itself off to one side of the cell, a little pocket forms around it, wrapped in membrane, and it may separate entirely, forming two curious "cells"—one with no nucleus and the other with practically nothing else but a nucleus. Neither one can survive for very long. The larger part, called a *cytoplast,* can live on for a few days, but it stops producing proteins after about twelve hours. The smaller portion,

containing the nucleus, is called a *karyoplast;* it continues to produce RNA for a few hours after it parts
from the original cell, and it can survive for about a
day.

Researchers at the University of Colorado in Boulder
took the process one step farther. They used Sendai
virus to fuse karyoplasts from one type of cell with
cytoplasts from another. They are working on combinations of mouse nuclei with hamster cytoplasts, to
see how well a hamster cell can translate the genetic
instructions from the mouse, and whether the hamster
cytoplasm can supply any genetic information that the
mouse genes do not contain. The unusual hybrid cells
should also be valuable in studying the mechanisms of
aging and cancer.

Since the mid-1970s, a whole new field of genetics
called recombinant DNA research has opened up.
Molecular biologists have devised methods of inserting
genetic information from other organisms into bacteria
and making them reproduce these foreign genes in huge
amounts. Researchers originally intended to use this
new technique to study the structure and functions of
genes, and it promises to be extremely useful for this
purpose. With large quantities of a gene available, the
task of determining its chemical structure becomes
much easier, and various aspects of its work can be
studied. (In the normal cell, many genes and their
products are present in such small amounts that they
can barely be detected, even by the most sensitive tests.)

As the new studies developed, researchers quickly
realized that recombinant DNA methods could also be
used to produce large quantities of valuable

biochemicals, perhaps even on a commercial scale. Visions of cheap and plentiful supplies of human insulin, growth hormone, interferon that could wipe out virus epidemics, Factor VIII for hemophiliacs, and a cornucopia of other useful products danced through their heads.

But visions of another sort began to haunt the minds of some recombinant DNA researchers. They saw the possibility of strange plagues against which we would have no defenses, of monstrous creatures the earth had never seen before, of an irreversible shifting of the delicate balance of life on our planet. Questions were asked, and answered, and argued. Soon a fiery debate was raging, first in the laboratories and then in the pages of newspapers and magazines. The recombinant DNA debate sparked a sweeping reexamination of science and scientists, and the degree to which lay people should be involved in directing their activities.

The Great Debate:
Recombinant DNA

What good is scientific research? Is the freedom to explore the secrets of nature one of the basic freedoms that should be defended in our society? Or are there some kinds of research that are just too dangerous to be allowed? Should restrictions be placed on certain investigations to protect the public from possible harm? Who should decide whether possible benefits are worth the possible risks? Scientists? They are the best qualified to understand and evaluate these benefits and risks—but they also have a strong interest, both emotional and financial, in continuing their work. The public? It seems reasonable that average citizens should have some say about activities that may affect their health and lives (and are supported by their tax dollars); but without the specialized knowledge and training of the scientists, can ordinary people have the understanding to make intelligent decisions?

These are hard questions. They are questions few people thought about, until public concern was crystallized by the great debate over recombinant DNA.

The great debate actually started back in the summer of 1971, although no one realized it at the time—and, indeed, the major recombinant DNA techniques had not yet been invented. Janet Mertz, a graduate student of Stanford University, was spending the summer at Cold Spring Harbor on Long Island, New York, taking a course on tumor viruses. She was particularly interested in the subject because her research group at Sanford, headed by Paul Berg, was studying a tumor virus called SV40. This virus is commonly found in monkeys (the letters *SV* stand for "simian virus"). It doesn't seem to cause monkeys any harm, and it also seems to be harmless to people. But SV40 causes cancer when it is injected into mice and hamsters; and when human cells in a tissue culture are treated with the virus, they change into a form very much like tumor cells.

Paul Berg wanted to find out why SV40 acts this way—which of the virus's three genes is responsible for causing cancer in small animals, and how this gene works. He had thought of inserting DNA from SV40 into a phage that normally lives in the bacterium *E. coli*. Janet Mertz mentioned Berg's plan to one of the lecturers at the summer course, a young cancer researcher named Robert Pollack. Pollack was already concerned about the way cancer viruses were being handled in cancer research laboratories. A number of scientists who had not been trained as microbiologists were beginning to work in the cancer field, and Pollack believed they often handled tumor viruses too carelessly.

When he heard about Berg's idea, Pollack was horrified. The Stanford researchers planned to put a tumor virus into a phage that naturally lives in *E. coli,* and *E. coli* naturally lives in people's intestines. What if a laboratory accident—or just sloppy handling—caused some people to be infected by the treated bacteria? Might they develop cancer? Pollack called Berg long distance to urge him not to do the experiment.

Imagine Berg's reaction after suddenly receiving a phone call from someone he had never even met, telling him he should stop his work because it was too dangerous. "At first it got my back up," Berg said later. But then he started to think about the experiment. He talked to other researchers and found that they had doubts and worries, too. No one really knew whether there was any danger, and if so, how much. Finally Berg decided to postpone the experiment until more information was obtained. He talked to other researchers, and plans to set up a study on whether people exposed to tumor viruses in the laboratory do eventually develop cancer were discussed.

In 1973, while scientists were beginning to think more and more about the safety of some of the new experiments in molecular genetics, a startling announcement was made. Two teams of researchers— Stanley Cohen and Annie Chang at Stanford University School of Medicine, and Herbert Boyer and Robert Helling at the University of California at San Francisco—published a joint paper. They had invented a new technique of gene splicing that made it possible to take a gene from any organism and insert it into the genetic information of bacteria. This was the recombinant DNA technique.

Like most scientific discoveries, the work of Cohen, Boyer, and their associates was based on experiments by other scientists who had gone before them. By 1973 it was known that the common bacterium *E. coli* has two kinds of genetic information. It has a single large chromosome made of DNA, the ends of which are joined together to form a ring. This ring of DNA contains about four million nucleotide pairs, enough to code more than a thousand genes. Many bacterial cells also contain various numbers of "minichromosomes," small rings of DNA that carry about a dozen genes. These minichromosomes are called *plasmids,* and some of them may be transferred from one bacterium to another during mating, carrying their hereditary information along with them. Resistance to certain antibiotics is one type of characteristic that is carried by some plasmids.

Earlier researchers had also discovered some interesting bacterial enzymes. One kind is called *restriction enzymes.* These proteins can cut a double strand of DNA, leaving two sticky ends dangling. Each restriction enzyme cuts DNA molecules only in certain places—wherever there is a particular sequence of nitrogen bases. For example, a restriction enzyme called Eco RI, produced by *E. coli,* cuts DNA between G and A wherever the nitrogen bases on one strand spell out GAATTC. The matching bases on the other strand of DNA are CTTAAG, which is GAATTC spelled backward. The enzyme cuts each strand between G and A, leaving two single-stranded ends, each spelled AATT.

Another useful type of enzyme is *DNA ligase,* which joins broken DNA chains together when the sticky ends are matched up.

The California research teams put all this information together into a simple technique for gene splicing. In plasmids they had a means of putting hereditary material into bacteria. Restriction enzymes were the "scissors" for snipping DNA into convenient-sized pieces for study, and also for opening up the plasmids so that new genes could be inserted. And DNA ligase was the "glue" needed to bind the DNA together into the new combination—the *recombinant* DNA.

This is how a typical recombinant DNA experiment works. First plasmids are isolated from bacteria. Then a restriction enzyme cuts them open, leaving dangling sticky ends. Meanwhile, the chromosome the researchers want to study is cut up with the same restriction enzyme. The DNA pieces will also have sticky ends, which just match those on the opened plasmid. The plasmids and DNA fragments are mixed together, and their sticky ends stick together. The breaks in the chains are mended with DNA ligase, and now there are new, larger plasmids, which have added a fragment of "foreign" DNA. The last step is to put the plasmids into bacteria. A soupy culture of *E. coli* is mixed with salt and plasmids, and then the flask with the broth is suddenly transferred from a dish of ice to a bath of warm water. The salt and the temperature change are a shock to the bacterial cells, causing the pores in their

RECOMBINANT DNA

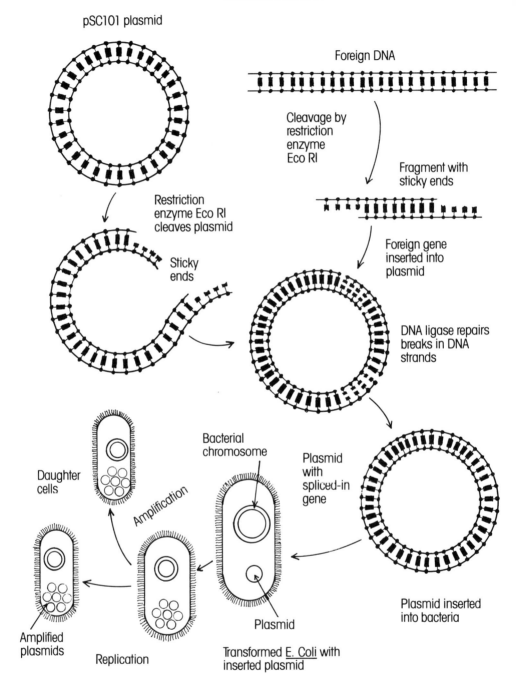

pSC101 plasmid

Foreign DNA

Cleavage by restriction enzyme Eco RI

Restriction enzyme Eco RI cleaves plasmid

Fragment with sticky ends

Sticky ends

Foreign gene inserted into plasmid

DNA ligase repairs breaks in DNA strands

Daughter cells

Amplification

Bacterial chromosome

Plasmid with spliced-in gene

Amplified plasmids

Replication

Plasmid

Transformed E. Coli with inserted plasmid

Plasmid inserted into bacteria

cell walls to widen, enough for plasmids to slip inside.

The bacteria containing the plasmids are grown on dishes of agar jelly. Each cell divides again and again, forming a clone and multiplying not only the bacterial chromosomes, but also the plasmids with their inserted DNA.

Molecular biologist Sydney Brenner later remarked, "the thing about recombinant DNA engineering is that it suddenly has made many things very easy that were once very difficult." Stanley Cohen and his associates quickly discovered this in their experiments with the new technique. First they tried transferring a gene from one type of bacterium to another. They took the gene for resistance to penicillin from a bacterium called *Staphylococcus aureus* (one of the genes whose location on the bacterial chromosome had been mapped), inserted it into a plasmid, and transferred it to *E. coli*. When the *E. coli* culture was treated with penicillin, the cells continued to grow. They had gained the drug resistance of the other bacterium.

E. coli and *Staphylococcus aureus* are not very closely related, but they are both bacteria. It might be expected that many of the processes that go on inside their cells

In recombinant DNA experiments, a plasmid is cut with restriction enzymes, and a "foreign" gene is spliced in with DNA ligase. The plasmid can then be amplified into thousands of copies, all with the potential for being translated into protein products.

would be rather similar; so it is not too surprising that a gene from one would work in the other. But could genes be successfully transferred over larger barriers—from a many-celled creature to a single-celled bacterium? Cohen's team decided to find out. They spliced genes from the South African clawed toad, *Xenopus laevis,* into plasmids and inserted them into *E. coli.* The genes were faithfully copied as the bacterial cells divided.

Cohen and Boyer's reports excited molecular biologists all over the world. Requests for supplies of the plasmids and enzymes began to flood in. Some of the proposed experiments were rather disturbing. When Cohen and Chang transferred the gene for penicillin resistance to their strain of *E. coli,* they carefully selected a type of drug resistance that was known to be commonly transferred in natural populations of bacteria. But some of the eager researchers who wanted to explore the new technique might not be so cautious. Some requests were from scientists who wanted to insert tumor virus genes into bacteria—just the kind of experiment that Paul Berg had decided was too dangerous to pursue. Some researchers proposed to use a technique called *shotgunning:* taking the entire chromosome set of an animal or plant—a fruit fly, for example—and chopping it up into fragments with restriction enzymes, separating the fragments, and then inserting each one into plasmids for study. Shotgunning would be a way to get large amounts of genes to study. But because these genes were largely unknown, the recombinant bacteria that would be produced would be unknown quantities. Some of them might turn out to have unexpected properties. And since cancer viruses are known to hide among the chromosomes of some animals, shotgun experiments might inadvertently introduce cancer genes into the laboratory bacteria.

In June 1973, just as these worries were beginning to be discussed, the annual Gordon Conference was held. The Gordon Conference is held each year in New England schools left empty for the summer vacation. Researchers from different laboratories get together and discuss their work. At the 1973 conference, Herbert Boyer gave a lecture on his recombinant DNA experiments. A lively discussion followed, and a few of the younger scientists at the meeting went to the cochairmen, Maxine Singer of the National Institutes of Health and Dieter Soll of Yale, to suggest a special session on possible safety problems in the research. The subject was squeezed into fifteen minutes on the last day of the conference. Fifty participants had already left, but the ninety remaining agreed to send a letter asking the National Academy of Sciences to appoint a committee to study possible health hazards of recombinant DNA research. Then another vote was taken, and by a majority of only six votes it was decided to send the letter to *Science,* one of the most widely read magazines for scientists.

The publication of the letter in the September 21 issue of *Science* transformed the debate from a quiet discussion among researchers in the field into a controversy that eventually involved all the major news media of the world. The public reacted with alarm: If the scientists themselves are worried, it must *really* be dangerous!

Meanwhile the scientists themselves, while beginning

to be somewhat appalled at the feverish public interest and suspicion, proceeded to study the problem. The National Academy of Sciences consulted Paul Berg, who scheduled a meeting with seven other researchers at MIT for April 1974. Meanwhile the California researchers, screening the requests that were flooding in for plasmids, were so alarmed at some of the proposed plasmid uses that they decided not to send out any plasmids until after the meeting.

The group at the MIT meeting quickly agreed that an international conference of scientists was needed to discuss the new research. But it takes time to organize a conference. One could not be held until the following February. What would happen in the meantime? One of the scientists at the conference remarked, "If we had any guts at all, we'd just tell people not to do these experiments. Maybe what we ought to do is make some public announcement."

That is just what they did. In a letter published in the journals *Science* and *Nature* in July 1974, the group of scientists asked researchers not to conduct certain types of experiments that might be risky: putting genes for antibiotic resistance or for the formation of toxins, like that of botulism, into bacteria that did not already have them; and inserting animal viruses into bacteria. The letter also advised that plans to put any genes from animals into viruses should be "weighed carefully," because of the possibility of accidentally introducing cancer genes.

Berg and his colleagues were asking the scientific world to put a moratorium on certain kinds of recombinant DNA research, to stop their work voluntarily, even though it seemed interesting and promising. The last time anything like that had been attempted was in the late 1930s, when a small group of atomic scientists, led by Leo Szilard, begged their colleagues to stop publishing information that might help the Germans to develop an atom bomb. They were practically ignored, until the government stepped in to clamp a lid of secrecy on all nuclear fission research. But the moratorium on recombinant DNA was carefully observed by scientists all over the world throughout the seven months that passed before the international conference was held at Asilomar.

In February 1975, about 150 scientists from all over the world gathered at the Asilomar Conference Center in Pacific Grove, California. Reporters also attended, some uninvited. The purpose of the conference was to discuss whether the new recombinant DNA research posed a health hazard to laboratory workers and to the public, and if so, what to do about it.

For three days the participants in the conference lived at Asilomar, completely immersed in discussions of recombinant DNA—attending meetings during the day and talking in small groups far into the night. Arguments raged between researchers who were most concerned about possible risks and those who were more concerned about the valuable time that had already been lost and wanted to get recombinant DNA studies on the way again as quickly as possible. A session on the last evening featured talks by four lawyers, who shocked the assembled scientists with the sober opinion that they had not only a moral but also a legal obligation: If they went ahead with their experiments without taking the most careful precautions and something went wrong—they could be sued.

Clearly the conference had to come up with some decisive result. The members of the organizing committee stayed up all night working on a statement that they hoped would be firm enough to show the public that scientists were acting on the highest principles, yet reasonable enough to satisfy the researchers who were eager to get back to work. The committee members never expected the large and varied group of researchers at the conference to agree on a single statement; they hoped to push their version through without a vote. Paul Berg presented the statement to the conference, and a vote was called for—so insistently that he had to agree. To the surprise of the bleary-eyed committee members, their recommendations were adopted almost unanimously.

The Asilomar conference recommended that certain recombinant DNA experiments should be conducted in special laboratories equipped with protective devices to ensure that no harmful products would infect laboratory workers or escape into the community. Work with animal viruses, for example, would be considered "moderate risk," and could be done only in laboratories with special air-filtering systems and safety cabinets, in which experimenters wearing gloves would handle the recombinant materials. "High-risk" work would be restricted to a few maximum-containment facilities, like the quarantine lab that was built to analyze the first samples of moon rocks brought home by astronauts before scientists knew they were not harboring any mysterious plagues from outer space.

Researchers working in such labs would enter them through air locks and would have to shower and change their clothing before leaving.

In addition to these plans for "physical containment," the conference called for the development of specially designed forms of bacteria, which could not live anywhere but in a laboratory dish and could not pass their plasmids to other bacteria if they happened to contaminate a human. The researchers at Asilomar optimistically believed that suitable bacteria for this "biological containment" could be developed within a few weeks. If they had realized how long it would really take, they might never have approved the resolution.

Starting the very day after the Asilomar conference, a committee set up by the National Institutes of Health (NIH) began work on developing a specific set of regulations for particular kinds of experiments, within the framework outlined by the conference. Four levels of increasingly strict physical containment were outlined and designated as P1, P2, P3, and P4. P1 was jokingly described as "working in an open lab out in your backyard garage," or the level of a "high-school biology lab," while P4 was so strict that there was only one laboratory in the whole United States that might qualify. Hundreds of thousands of dollars would have to be spent to modify existing laboratories to the requirements of P3 facilities.

Three levels of biological containment were planned: EK1 (the standard laboratory strain *E. coli* K12), EK2 (a specially weakened strain), and EK3 (a weakened strain that had passed stringent tests to show that it was *really* safe). Roy Curtiss of the University of Alabama eventually developed a specially disabled bacterium, which he named chi-1776 in honor of the Bicentennial, but it was not approved even for the EK2 level until December 1976, nearly two years after the Asilomar conference. The NIH, meanwhile, took sixteen months to work out its guidelines for recombinant DNA research, which would apply to all institutions receiving federal funding for their work. Most other research laboratories agreed to follow the guidelines voluntarily, and other countries around the world devised their own codes, basically patterned after the NIH version.

While the regulations were being worked out, some recombinant DNA research continued, but many of the most interesting and promising experiments still had to be suspended until the proper laboratory facilities and specially weakened bacteria were available.

Anyone who had hoped that the announcement of the NIH guidelines would settle the recombinant DNA problems once and for all was sadly mistaken. The debate raged on. Congress began to hold hearings on the possible need for laws to regulate the new research. The news in 1976 that Harvard University was quietly planning to convert a laboratory to a P3 facility without asking for special permission stung the mayor of Cambridge, Massachusetts, to a fury. Mayor Alfred Vellucci called a public hearing on the matter, which turned into a televised circus. Researchers Maxine Singer and Mark Ptashne, a Harvard biologist, contended that the proposed research was safe and the lab would pose no danger to the community. They were angrily debated by equally prestigious scientists, including Nobel Prize winner George Wald and MIT

Work in a P4 laboratory at Ft. Detrick: experiments are carried out in air-tight stainless steel cabinets by workers using shoulder-length heavy gloves that are tightly attached to the openings in the cabinet.

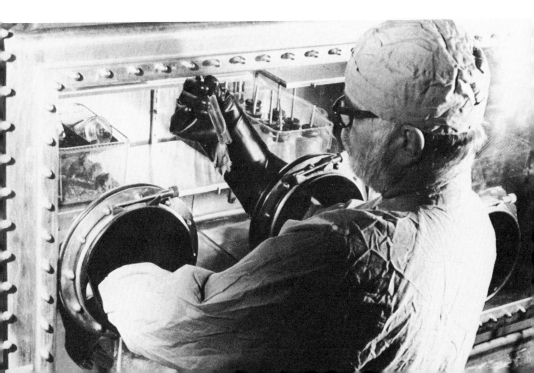

researcher Jonathan King, a prominent member of a radical group called Science for the People.

Mayor Vellucci called for a two-year ban on all recombinant DNA research in Cambridge (which would have covered both Harvard University and MIT); the city council eventually voted on a three-month moratorium, which was later extended while a committee of doctors and lay people considered the problem. The nine Cambridge citizens on the committee were under a worldwide spotlight, and they worked diligently. They met twice a week to hear testimony on the safety of recombinant DNA experiments and carefully studied the scientific principles involved. Eventually they recommended that the recombinant DNA research be allowed to continue, and the city council approved their recommendation.

What was all the sound and fury about? What did the opponents of recombinant DNA research fear? Were their fears justified?

The choice of *E. coli* for the experiments raised the greatest concern. In a way, it was an obvious choice. This bacterium is one of the most widely studied. Many of its genes have been mapped, and about a third of its chemical reactions are known. So in introducing unknown genes into *E. coli* scientists would be working with only one mystery, not two. But *E. coli* is a microbe that normally lives in people's intestines. Worried researchers speculated on a long list of possible scenarios, all starting with "What if *E. coli* with recombinant genes escaped and infected someone?" The inserted genes might turn the bacterium into a disease germ, which might be resistant to all known antibiotics. If tumor genes were introduced, on purpose or accidentally, the recombinant bacterium might cause cancer. If inserted genes from animals were translated into proteins, they might stimulate the body's defenses to form antibodies against the foreign proteins; but these proteins might be similar enough to human proteins so that the antibodies would attack the person's own body chemicals, too. Recombinant bacteria specially bred to produce insulin or other hormones on an industrial scale might form these hormones in the body and cause illness by upsetting the natural hormone balance.

Geneticist Robert Sinsheimer, a pioneer in genetic engineering, was a vocal opponent of the new research. Some years before, Sinsheimer had argued persuasively against scientists who felt that research on genetic engineering should be deliberately slowed down until we gain the wisdom to use the new knowledge properly. "I would suggest," he said, "that those who feel this way are not among the losers in that chromosomal lottery that so firmly channels our human destinies. This response does not come from the 250,000 children born each year in this country with structural or functional defects." But now he was having second thoughts and raised a subtle philosophical point. Through millions of years until now, Sinsheimer said, nature has had built-in genetic barriers between species, preventing the exchange of hereditary information. Now, with recombinant DNA techniques, we are breaking down these evolutionary barriers and rewriting the rules of the game. And no one can predict what effect this will have on our world and its delicate balance of life. Other

researchers countered this argument with the statement
that the barriers are imaginary, that exchanges of information between bacteria and the higher organisms
they infect go on all the time. We are not breaking the
rules of evolution, they claim, because the rule is that
"anything goes."

During the past few years, these other fears have been
gradually receding. One of the most encouraging findings arose out of a problem that the recombinant DNA
researchers encountered. Much of our knowledge about
the workings of DNA and RNA has been gained in work
on microorganisms, and it had been assumed that their
genes and those of higher organisms work in the same
way. But nobody could know for sure. When researchers inserted genes from animals and plants into bacteria,
the microbes faithfully copied the DNA as they divided.
Ways were even found to *amplify* the genes—to trick
the bacteria into multiplying the plasmids without
dividing, so that a single cell would contain hundreds of
copies of a plasmid with its inserted gene. It did not
prove too difficult to go to the next step and show that
the DNA was transcribed into RNA copies. But then the
researchers hit a stone wall. The messenger RNAs
produced from recombinant genes of higher organisms wouldn't translate into protein in the

*In mammalian genes the genetic information, spelled out in
nucleotide code, is broken up by intervening sequences of
nucleotides that are not part of the gene's message. Mammalian cells "process" mRNA to remove the intervening
sequences before translation into proteins, but bacterial cells
do not have the ability to do this.*

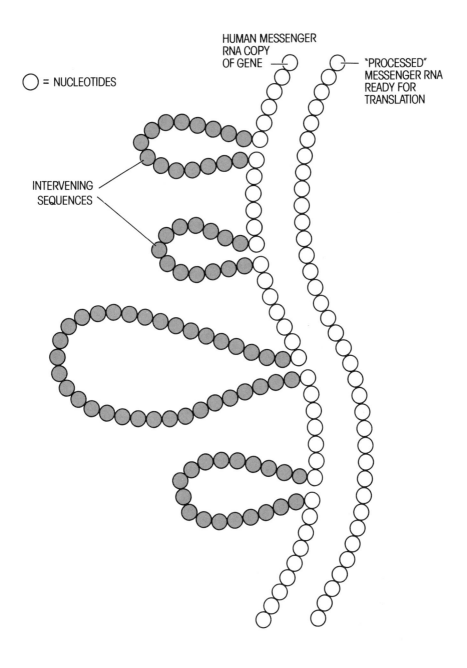

HUMAN MESSENGER
RNA COPY
OF GENE

"PROCESSED"
MESSENGER RNA
READY FOR
TRANSLATION

◯ = NUCLEOTIDES

INTERVENING
SEQUENCES

bacterial cells. Or if they did form protein, the proteins didn't have the structure they were supposed to.

Eventually the difficulties were identified, and the problems were solved. One stumbling block turned out to be the fact that bacteria and higher organisms use different kinds of "switches" to turn their genes on and off. Researchers learned to add the appropriate switches when they were working with animal genes. (One trick is to attach the gene to a gene for a protein the bacterium normally makes anyway. Then a large hybrid protein is obtained and split apart to get the desired animal protein.)

But in tracking down a second stumbling block, molecular biologists made a major new discovery. A bacterial gene is typically a chain of DNA with the coded message for a protein structure spelled out in sequence. But genes of humans and other animals turn out to be more complicated. First comes part of the message. Then there is a stretch of DNA that doesn't match the amino acid sequence on the protein; in fact, it doesn't seem to mean anything that scientists have yet discovered. Then comes another part of the message, then another stretch of "meaningless" DNA, and so on. In an animal cell, the DNA is copied in RNA, and then the meaningless *intervening sequences,* or *introns*, are removed by special enzymes, to form the working messenger RNA that carries the plan for a protein. Bacteria don't have the right enzyme systems to cope with the introns in animal genes; so they can't translate them into proteins.

Once the problem was identified, researchers quickly found ways around this stumbling block as well. They could get translatable genes by synthesizing them according to Khorana's "sticky-end" method, or by working backward from messenger RNA using reverse transcriptase. But the discovery that bacteria *can't* translate animal genes they may pick up means that researchers can use recombinant microbes as a tool for studying animal genes without fearing that they will cause trouble.

Experience in recombinant DNA work so far has repeatedly shown these experiments to be much safer than scientists expected. The NIH has been gradually relaxing its guidelines, classifying many types of experiments as less risky and permitting them to be conducted under much less strict conditions. In addition, *E. coli* K12 itself, even without the special weakening, has turned out to be a rather finicky and delicate creature, unlikely to survive if it escapes from its laboratory dish.

When the debate was at its hottest, opponents of recombinant DNA research criticized those who were willing to hazard possible risks for "theoretical benefits." But practical experience is showing that it is the risks that are "theoretical," while the practical benefits are already starting to appear.

A library of cloned human genes being prepared by T. Maniatis at Cal Tech is providing researchers with the materials they need to study our genetic heritage.

Early in 1978 Stanley Cohen and his associates announced the successful translation of a synthetic gene for the hormone somatostatin. A commercial firm is setting up the manufacture of this hormone, which will be a useful research tool for learning about the en-

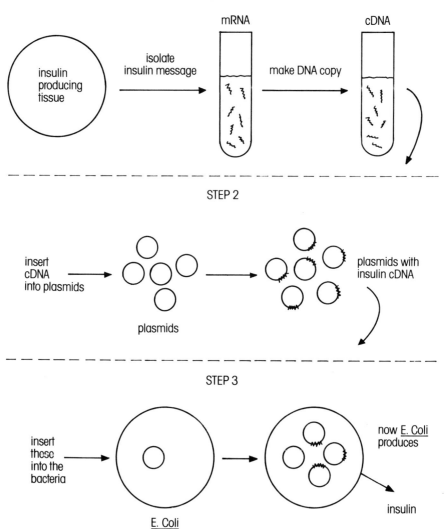

The hormone insulin has been produced by recombinant DNA techniques.

docrine system and may also be valuable in treating some growth disorders and even diabetes.

In June, 1978, a team of researchers at Harvard University and the Joslin Diabetes Foundation in Boston, led by Walter Gilbert, announced that they had successfully used recombinant DNA techniques to synthesize a form of rat insulin. Only a few months later, researchers at the City of Hope Medical Center in

This scanning electron micrograph shows the bacteria used by University of California researchers to produce rat insulin by recombinant DNA techniques.

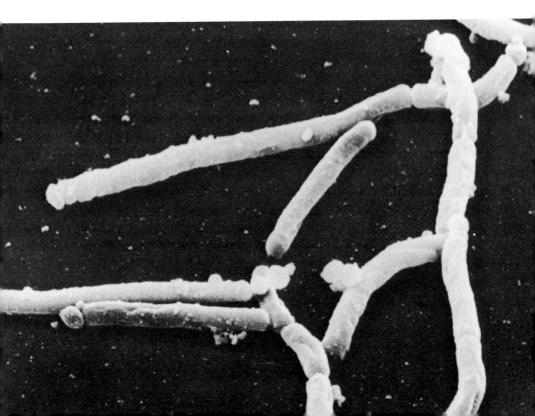

Duarte, California, announced that they had produced *human* insulin in a similar way.

Drug companies have actively entered the recombinant DNA field, and announcements of valuable new products are starting to come in—so rapidly that almost anything else we can say in this chapter will probably be out of date by the time you read it. For example, human growth hormone can be used to help very short people reach a normal height. So far this hormone has been produced very laboriously by extracting it from pituitary glands of dead people. One hundred fifty pituitary glands are needed to get enough growth hormone to treat just one child for a year. But now human growth hormone has been produced in *E. coli*. By the time you read these words, the hormone may be in production by a drug company.

Early in 1980, while this book was already at the printer, an international team of recombinant DNA researchers made another exciting announcement: They had used *E. coli* to synthesize interferon. This protein had been at the top of molecular biologists' "shopping list" for a long time. It is a key part of the body's defenses against virus infections, such as measles, chicken pox, colds, and flu. Experiments with interferon laboriously extracted in tiny amounts from human blood cells have also shown promise for this chemical in the treatment of cancer. A single course of treatment with the extracted interferon cost as much as $50,000, but recombinant DNA synthesis will soon be providing cheap new sources of the protein.

Tinkering with Genes: Genetic Engineering

Recombinant DNA research, moving along briskly now that much of the debate has faded away, is providing a wealth of new knowledge and many valuable products. But recombinant DNA is not the whole story of genetic engineering, by any means. Even while the recombinant DNA studies were stalled, waiting for the risks to be evaluated and guidelines to be established, geneticists and molecular biologists were pursuing a number of other promising lines of research. Some, like recombinant DNA techniques, promise to place a cornucopia of biochemicals within our grasp. Others may make it possible to go right into the heart of cells and modify their heredity.

Biologists are buzzing these days about a new kind of laboratory creation called the *hybridoma*. It all started in the mid-1970s, when Cesar Milstein and his colleagues at the Medical Research Council Laboratory of Molecular Biology in Cambridge, England, were working on a rather theoretical problem. They were studying antibodies, the large and complicated proteins produced by certain cells when the body is invaded by a virus, bacterium, or some foreign chemical. Milstein was curious about whether parts from different antibody molecules could be put together to form a new kind of hybrid antibody. He decided to find out by using cell fusion to join two types of antibody-producing cells: mouse spleen cells and cells of a fast-growing cancer called mouse myeloma. In the experiments he started out by injecting an antigen—a foreign substance such as a chemical, a virus, or a bit of tissue—into a mouse. The animal promptly produced antibodies that attacked the antigen. There was nothing new about this part of the experiment; this is the standard way of obtaining antisera such as tetanus toxoid to provide temporary protection against diseases. (Usually a much larger animal than a mouse is used.) But Milstein did not collect the antibodies from the animal's blood; instead he collected the spleen cells that produced the antibodies. Spleen cells normally do not grow very well in cultures. Myeloma cells, on the other hand, are hardy cells that grow very rapidly, not only in the body but also in a culture dish. The combination of spleen cells fused with myeloma cells also grew well.

Milstein was pleased to discover that his original idea was correct: the hybrid cells—"hybridomas"— produced some hybrid antibodies that combined features of the spleen-cell antibodies and the antibodies naturally produced by the myeloma cells. They also produced a mishmash of other antibodies. Some of the hybridomas were quite specialized, producing antibodies only against a particular antigen. By selecting these specialist cells and cloning them—growing whole cultures from a single parent cell—Milstein and his colleagues were able to obtain large quantities of almost pure antibodies, which they call *monoclonal antibodies.*

The new technique provides a way to obtain unlimited amounts of antibodies against any virus, bacterium, or tissue cell, as well as chemicals such as proteins and carbohydrates. In addition to their obvious use in treating viral and bacterial diseases, these antibodies have an amazing variety of possible applications. Researchers are starting to use them to examine the molecules on the outer surfaces of cells. When cells are injected into a mouse and then its spleen cells are converted to hybridomas, a variety of antibodies will be obtained, each matching a particular part of the cell surface. Thus, hybridomas are a tool for mapping the cell surface. They can provide information about the important reactions that take place at the contacts between cells, and perhaps even reveal some key differences between normal cells and cancer cells that may suggest new treatments for cancer.

Antibodies produced by hybridomas may themselves be used as weapons against cancer. In one recent experiment, hybridomas producing antibodies against a deadly form of cancer called melanoma were injected into mice. Later, melanoma cells were injected into the mice. Normally they would form tumors, but the mice protected with hybridomas remained cancer-free.

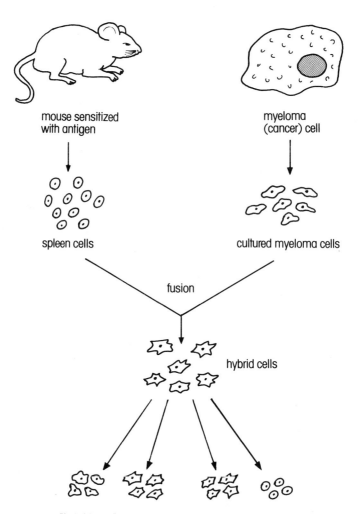

mouse sensitized
with antigen

myeloma
(cancer) cell

spleen cells

cultured myeloma cells

fusion

hybrid cells

"hybridoma" cell clones producing specific antibodies

Cell fusion combining antibody-producing spleen cells with fast-growing tumor cells yields hybridomas that are a valuable source of monoclonal antibodies. Similar techniques may be used to produce other biochemicals.

Hybridomas producing antibodies against viruses are being used to study disease germs and also in treatments to combat disease. Monoclonal antibodies produced against rabies virus have already been shown to protect mice from this deadly microbe. Another team of researchers has produced human tetanus toxoid antibodies using hybridomas.

As recombinant DNA research expands, an important use for monoclonal antibodies will be in picking out protein products manufactured by the recombinant bacteria. Shotgunning, for example, could provide much valuable information about genes if the bacteria that had incorporated each fraction of the chromosomes could be neatly separated. But a typical shotgun experiment produces a seemingly hopeless muddle of products that could never be sorted out by ordinary chemical means. Monoclonal antibodies will provide fine-tuned probes, which can reach into the mixture and identify bacteria making a particular protein. "Give me any antibody directed against a protein," says Swiss researcher Bernard Mach, "and with time that gene can be cloned."

An even more exciting possibility is that hybridomas themselves could be used to make not only antibodies, but practically any other cell product in huge quantities. This could be done by starting with cells that specialize in producing a particular substance—for example, pituitary cells that make growth hormone—instead of spleen cells.

Cultures of plant cells are providing another source of useful biochemicals. Drug researchers in England are now setting up the manufacture of birth control pills from ingredients produced in huge vats of plant cell cultures. (The hormones used in the pills are normally extracted from yams, grown in fields in Mexico.) Laboratories in West Germany and Japan have used plant cell cultures to produce fodder for silkworms and tobacco. Morphine, digitalis, and other drugs normally extracted from plants, as well as substances used as flavorings, food additives, and laxatives, could also be produced in this way. Researchers have found they can greatly increase the natural productivity of the plant materials by selecting and cloning the single cells that give the highest yields of the product.

Large-scale culturing of microbes such as bacteria and the mold that produces penicillin has been a big business for many years. But the technical problems involved in culturing plant cells have just recently been solved. Now that plant cultures are becoming more common, they will provide another medium for the work of genetic engineers. Genes for useful products might be introduced into plant cells by attaching them to viruses that normally infect the plants.

Recombinant DNA, hybridomas, and other cell-culturing techniques will soon be providing physicians with supplies of biochemicals that can be used to treat a number of genetic diseases. In many metabolic diseases, for example, the symptoms are the result of the lack of some key enzyme. If this enzyme could be replaced, the condition could be corrected.

Picture a teen-ager living under a death sentence. Constant pains in his arms and legs remind him that he is doomed to die by his early forties. No earthly judge has passed the death sentence on him. His only "crime" was to be born missing a particular enzyme of fat metabolism. Without this key enzyme, fats are slowly building up in his kidneys, clogging them and making them ever less able to rid his body of the waste products that are poisoning him. He suffers from a hereditary condition called Fabry's disease. His only hope of a reprieve lies in medical progress.

The hope is real. Researcher Roscoe Brady and his NIH team have taken the first steps in replacing the lacking enzymes in patients with defects of fat metabolism. The first results have been promising, but so far the treatment can be used only in brief experiments. There just isn't enough of the enzymes to go around. Recombinant DNA research can change this picture and make enzyme replacement a routine treatment.

Tay-Sachs disease, the tragic wasting sickness that kills young children, is another disorder of fat metabolism. The Brady team has been working on this, too, but so far without much success, even though they know exactly which enzyme is lacking and have tried to introduce it. The stumbling block is that the damage in Tay-Sachs disease is caused by a buildup of fats inside the brain cells, and the body has formidable defenses protecting the brain from invasion by foreign substances—even the enzymes that might help Tay-Sachs patients. New techniques will be needed to transport the enzymes to where they are needed.

The barrier that protects the brain is only one of many obstacles that continually frustrate medical researchers in their efforts to correct the body's errors by introducing biochemicals. Substances taken by mouth are often digested before they can do their work. Insulin, for example, cannot be taken in pills but must be injected; as a protein, it would be quickly broken down in the stomach. Enzymes, too, are proteins. Even injection of these into the tissues or bloodstream does not always guarantee success since proteins and other biochemicals may be recognized by the body's defenders as "foreign" substances and so trigger an attack by antibodies. Even if they escape this hazard, the biochemicals still may not be able to enter the cells where they must do their work. Or they may break down quickly and have to be continually replaced.

Researchers are working on a number of ways around these difficulties. One of the most ingenious solutions is to sneak biochemicals past the body's defenses by wrapping them up in tiny packages called *microcapsules*. These have been made of various materials, from plasticlike polymers to liquid films. Some of the microcapsules are very much like the membrane of cells.

Microcapsules may be sievelike, with their membranes pierced by tiny holes too small for the biomolecules inside them to pass through. Safe in their protective custody, introduced enzymes are shielded from attack by the body's defenders, but they can still

Microcapsules hold enzymes, drugs, or other chemicals inside a tiny envelope. Their membrane may permit small molecules to pass in and out of the microcapsule; or the membrane may be gradually broken down by cell enzymes, releasing the introduced chemical inside the cell.

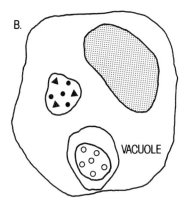

CELL MEMBRANE

A.

LYSOSOME   NUCLEUS

ENZYME CONTAINING
MICROCAPSULE

B.

VACUOLE

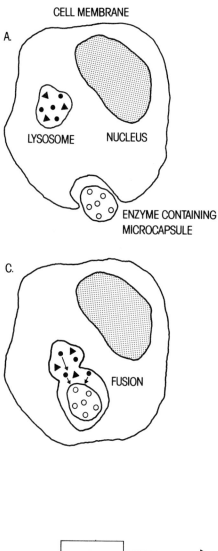

C.

FUSION

D.

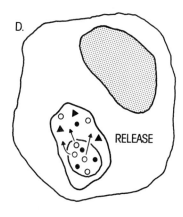

RELEASE

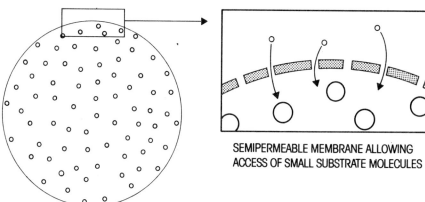

SEMIPERMEABLE MEMBRANE ALLOWING
ACCESS OF SMALL SUBSTRATE MOLECULES

MICROCAPSULE

do their job. The tiny pores are large enough to let small body chemicals inside, to be worked on by the enzymes.

Some microcapsules are biodegradable. Like tiny time capsules, they are gradually broken down by the body, releasing their active chemicals bit by bit.

Injected microcapsules generally wind up mainly in the liver and spleen. But researchers are now developing ways to attach tiny "tags" that target them for pickup by other body organs. Attaching antibodies, for example, may make white blood cells eagerly gobble up the microcapsules. These roving cells travel through the blood and tissues, and they can dispense doses of a drug or enzyme as they go.

Treating a disease like diabetes or one of the enzyme-defect disorders by replacing the missing biochemical is a lifetime affair. A person must take the medicine every day, forever. How much better it would be to correct what is wrong with the body, once and for all. Genetic engineers hope that their research will make it possible to do just this: to repair or replace the genes that don't work properly, so that the body can carry on without the need for further treatment.

Some people have criticized genetic engineering research and even called for a stop to it. They say that it is wrong, because it will change people into something unnatural. Think about it, reply the genetics researchers. People die "naturally" from cancer and heart disease. Each year, in the "natural" course of events, a quarter of a million American babies are born physically or mentally crippled. There is so much suffering. If there is a way to help, how can we say, "No, we will not do it"?

One way to correct a faulty gene is actually to change the DNA on the chromosomes chemically. Molecular biologists have taken the first steps toward an ability to do this. Researchers at the City of Hope National Medical Center in Duarte, California, working with the tiny phage φX174, were able to repair a mutation consisting of one change in a single nucleotide in its DNA. They did it by adding a short chain of nucleotides spelling out the correct message. The synthetic molecule paired up with a portion of the phage DNA (except in the one spot where a nucleotide had mutated), and enzymes were used to make the phage complete the ring; then the new strand was used as a pattern for the formation of a new phage chromosome with the chemical error corrected. Of course, it is a long way from φX174 to a human cell, and it will probably be a long time before scientists can zero in on particular cells in the body and repair specific spots on their DNA. A solution within much closer reach will be the insertion of extra genes, perhaps carried into the cell by viruses.

The idea of using viruses to carry genes piggyback into cells came from observing what happens in nature. A process called *transduction* was first described by bacteriologists back in 1952. Phages, the viruses that infect bacteria, can be transmitted from one bacterium to another, much as flu viruses are passed from one person to another. In the process, the phages can carry hereditary information with them. In the first bacterium the phage may pick up bacterial genes, which become attached to the phage nucleic acid. Then, when such phages infect another bacterial cell, they bring their baggage of bacterial genes along with them, and the genes from the first bacterium may be added to the genome of the second.

In the late 1960s an NIMH medical researcher, Carl Merril, was taking a course on phages at Cold Spring Harbor Laboratory. He was surprised to see how casually the phages were handled. Samples were drawn up into pipettes held in the mouth; an experimenter who sucked a little too hard wound up with a mouthful of culture.

Everybody knew that phages infect bacteria. But Merril wondered if these viruses might also have an effect on humans. He started a study of a virus called lambda phage. This phage is known to carry some genes for the metabolism of a sugar called galactose, which is found in milk. Merril combined the phage with cultured cells from human patients with galactosemia. People with this genetic disease lack the enzyme for using galactose, so they cannot gain nourishment from milk. Galactose builds up in their bodies in such large amounts that it becomes a poison. Babies with galactosemia suffer from liver diseases, cataracts, and mental retardation.

Merril had a theory: If phages could infect human cells, in addition to bacteria, then the lambda phage would supply the genes for the missing enzyme to the human cells in the culture. Sure enough, that is exactly what happened. The galactosemic cells were "cured," at least in a test tube, and they passed on the ability to produce the enzyme to later cell generations.

Meanwhile, other researchers were investigating the possibility of using viruses that normally infect animal cells to transfer genes. Such studies are potentially dangerous, since many viruses carry diseases. Virus carriers for genetic engineering must be very carefully selected.

The first experiment started by accident. More than forty years ago it was discovered that warts on the skin of cottontail rabbits in western Kansas are caused by a virus called Shope papilloma virus. The virus does not seem to produce any disease in humans (not even warts). In fact, its discoverer, Richard Shope, was so convinced it was safe that he injected himself with the virus in 1933; he reported that nothing happened.

In 1959 biochemist Stanfield Rogers discovered that something does happen to people exposed to Shope papilloma virus. He had shown that the virus causes the production of arginase, an enzyme that breaks down the amino acid arginine. In rabbits infected with the virus, the warts produce large amounts of arginase, and the blood of the rabbits contains unusually low amounts of arginine. The virus enzyme is working in their bodies.

Rogers began a follow-up study of about seventy-five laboratory workers who had been exposed to the Shope papilloma virus during the past decades. He found that *all* of them had unusually low levels of arginine in their blood. They were all carrying the virus gene. All of them were healthy. Neither the virus nor the low levels of the amino acid seemed to be causing any problems. Rogers concluded that he had discovered "a form of treatment with no known disease."

Since genetic diseases often turn out to be due to a lacking enzyme, it was just a matter of time before a matching disease—arginemia—turned up. In 1969 geneticist Joshua Lederberg of Stanford University sent

Rogers a note calling his attention to two young sisters in Germany with a genetic disease involving a lack of arginase. Their blood serum arginine levels were ten times higher than normal, and they were severely mentally retarded. Rogers wrote to the German doctors who were treating the case, offering to "challenge" the new disease with Shope papilloma virus. Small skin samples from both girls were flown to Rogers's laboratory in Oak Ridge, where they were cultured and infected with purified virus. Sure enough, the virus genes began to produce arginase in the girls' cells. Meanwhile, however, the girls were getting worse. Rogers sent some samples of virus to Germany, and the doctors tried to treat the young victims of the disease with virus injections. Their serum arginine levels were somewhat lowered, but the girls remained mentally retarded. Apparently it was too late to repair the brain damage.

Then the parents decided to ignore the advice of the genetic counselors and had another baby girl in 1971. She, too, was soon diagnosed as suffering from arginemia. A single dose of the virus was given to the baby when she was a few months old, but she did not improve. A hail of medical criticism rained down on Rogers for using an untested treatment on human patients. (If it had worked, he would have been hailed as a hero.) Rogers argued that forty years of experimentation had already shown that the virus was safe, and since there was no other treatment for the disease it was worth the chance—the sisters were doomed otherwise. Eventually it was discovered that the treatment of the third child was not a real test. The viruses in the dose she received had died in the mail on the way to Germany, so she never received the live virus at all.

Someday the experiment will be tried again. Meanwhile, other genetic engineers are taking the more conventional route of working up from test-tube studies in the laboratory. Late in 1978 Stanford University researchers announced that they had successfully transferred a gene from one type of mammalian cell to cells of another species, using the SV40 virus. Using restriction enzymes, a team headed by Paul Berg cut out the portion of the viral DNA that codes for a protein in the outer coat of the virus. Then they inserted the rabbit gene for hemoglobin at this point. This step accomplished several goals. It made the virus unable to reproduce (since one of its key capsule proteins was missing), placed the new gene in a position where it could be read properly by the cell's protein-making systems, and left the virus about the same size and shape as it had been originally. Then the altered virus infected a culture of African green monkey cells. Tests showed that the monkey cells began producing rabbit hemoglobin.

The researchers expect that Berg's accomplishment will be most immediately useful as a tool in gene-mapping and in studying how genes work. But a modification of the technique might one day be used as a cure for sickle-cell anemia and other diseases due to abnormal hemoglobin. The gene for normal human hemoglobin could be inserted into a virus (perhaps SV40, perhaps another, depending on whether it is definitely shown to be safe). Then bone marrow cells

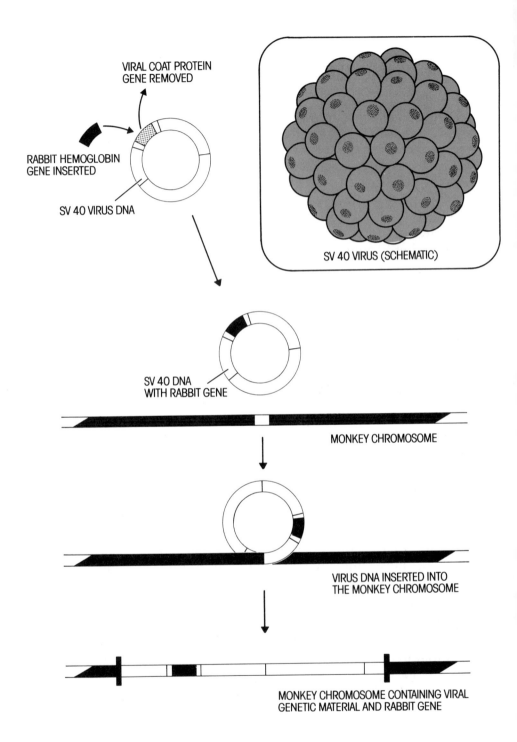

VIRAL COAT PROTEIN
GENE REMOVED

RABBIT HEMOGLOBIN
GENE INSERTED

SV 40 VIRUS DNA

SV 40 VIRUS (SCHEMATIC)

SV 40 DNA
WITH RABBIT GENE

MONKEY CHROMOSOME

VIRUS DNA INSERTED INTO
THE MONKEY CHROMOSOME

MONKEY CHROMOSOME CONTAINING VIRAL
GENETIC MATERIAL AND RABBIT GENE

from a sickle-cell anemia patient could be removed, infected with the altered virus, multiplied in culture, and finally reinjected into the bone marrow. All the red blood cells produced from these "re-engineered" marrow cells would produce some hemoglobin S (coded by the old gene) and some normal hemoglobin (coded by the gene carried by the virus). Then the patient's blood would be similar to that of a carrier of the sickle-cell trait, and the cells would no longer tend to sickle.

Virus transfer is one way to introduce new genes into cells. Another that may be used by future genetic engineers is cell fusion. This may someday be a standard procedure for treating genetic diseases: Take a sample of the patient's cells—the kind that need the lacking enzyme to do their job in the body. Then fuse them with cells (perhaps mouse cells) that produce this substance. Search through the numerous cells growing in the fused culture for hybrids that have the right activity. The cells must also pass a test for compatibility with the person's own cells; otherwise they will be attacked by antibodies when they are reinjected. Finally, reinject the cells into the patient's body to multiply and produce the missing chemical.

Researchers have used the SV 40 virus to carry a rabbit hemoglobin gene into monkey cells, where it was incorporated into the monkey chromosome and produced its protein product.

We can't do all that yet, but some promising experiments suggest that such an approach would work. One such experiment was reported late in 1973. A team of researchers took white blood cells from mice with a genetic defect: They could not produce a key protein needed for antibody reactions. The cells were fused with kidney cells from normal mice. Kidney cells normally do not make the missing protein, but their chromosome set contains the gene for it. In the hybrid cells the kidney-cell genes were turned on, and the cells began to produce the missing protein! They were injected into the mice and gave them a normal antibody system.

The mice did not stay cured, though. After a time they started making antibodies against the new protein. The researchers believe that this would not be a problem in most genetic diseases. The immunity system of the mice fought the protein because it was completely foreign—before the hybrid cells were injected, their bodies had never produced any of the protein. In most metabolic diseases, however, the problem is merely that the gene is faulty; it produces a little of an enzyme or some other key chemical, but not enough to keep the body working properly. So injected hybrid cells would just be adding more of the same protein, not introducing a new one.

Even if the protein product is not a "foreign" chemical, the hybrid cells themselves may be. The cell that contributed the gene for the missing protein may also contribute some alien surface chemicals to the hybrid. This is all the more likely in mouse-human hybrids, which tend to lose human chromosomes in preference to mouse chromosomes.

Researchers have already figured out two possible ways around this problem. One is to use red blood cells as the gene donors. When such cells form hybrids, they do not contribute any of their membranes to the surface of the "new" cell; so the body will still recognize the hybrid cells as its own. Another approach is to supply the missing genes with sperm cells. A sperm is a tadpolelike structure, with a long tail that helps it to "swim" and a head that is basically a packet of chromosomes. Researchers at Sloan-Kettering mixed mouse sperm cells with hamster fibroblasts (a kind of skin cell). The sperm burrowed into the cells just as though they were fertilizing an egg. Some of the fused cells produced mouse proteins. In later studies, Columbia University researchers used normal mouse sperm to supply a missing enzyme to mouse fibroblasts. Only about one out of every million cell clones obtained from the fusion produced the missing enzyme—but this was enough, since the cell clones could quickly be multiplied to huge numbers.

Late in 1979, researchers from the National Heart, Lung, and Blood Institute and Rockefeller University announced the development of a powerful new technique for introducing genes into cells. Working under a microscope, they squirted tiny bits of material containing single genes directly into individual cells through a hollow glass needle only one-thousandth of a millimeter in diameter. In one experiment, a virus gene for producing an enzyme called thymidine kinase was injected into mouse cells lacking the ability to make that enzyme. Such genetically crippled cells normally cannot survive, but after the delicate genetic surgery, the cells thrived and multiplied in their culture dishes. In another experiment, the genetic engineers inserted a human gene for beta globin, a portion of the hemoglobin molecule. In addition to the possibility of curing genetic diseases by modifying cells' heredity, this exciting new technique will also make it possible to study the way cells live and work: Researchers can insert a normal gene into one cell and an abnormal gene into another similar cell and observe the results.

It seems that we are well on the way to an ability to correct nature's errors by tinkering with the genes themselves. But genetic engineering may also make possible even bolder undertakings: Someday soon, we may be able to try our hand at improving on nature.

Improving on Nature

In 1978 two sensational events excited the public's imagination and focused attention on the new trends in genetics research.

The first was the publication of a book, *In His Image,* by David Rorvik. The book was published as a work of nonfiction and claimed to be the story of the first cloning of a human being—a child who was an exact genetic copy of his father. Scientists are generally agreed that the book was actually fiction. Someday we may indeed have the ability to produce exact copies of ourselves, but research has not yet progressed that far.

The second event sounded just as science-fictional, but it was fact: the birth of the first "test-tube baby." Baby Louise spent most of her life before birth growing in the usual way, inside the body of her mother. But she was conceived in a mixture of egg and sperm in a laboratory culture dish. The birth of Baby Louise did

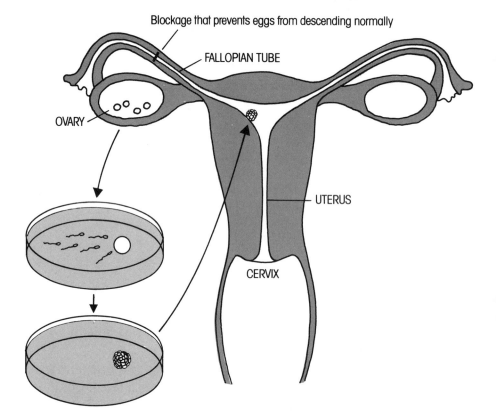

Conception of a "test-tube baby." First the woman is treated
with hormones that stimulate eggs in the ovary to mature.
Using an instrument called a laparoscope, the surgeon locates
the ovary and withdraws the eggs with a hollow needle. An
egg is placed in a culture dish with blood serum and nutrients
and fertilized with the sperm. The fertilized egg is grown in
culture for three to six days, dividing to form a blastocyst.
The blastocyst is placed in the woman's uterus, after she has
received further hormone treatments, and it develops normally into a baby.

not involve any actual gene manipulations: A 23-chromosome assortment from her mother joined with a set from her father in much the same way as if she had been conceived in a more conventional manner. But the techniques developed by the successful British researchers Robert Edwards and Patrick Steptoe to fertilize the egg, culture it through the first stages of development, and then implant the embryo back into the uterus of its mother, are important first steps in gaining the skills researchers will need if they do make serious attempts to modify the genes of children before birth.

The ideas of cloning people and manipulating genes are not new. *Brave New World,* the famous novel written by Aldous Huxley in 1932, shocked people with its speculations on producing endless rows of identical babies, specially designed for particular roles in society and raised in artificial wombs from which they are "decanted" rather than born. Most people are repelled and upset by the idea of a society as rigid and ruthless as that depicted in *Brave New World* and of the carbon-copy drudges who would make it possible. But some speculations have provided a more positive view of cloning, including a perpetuation of the genes of great statesmen, artists, and thinkers. (The clones, however, might not grow up to be statesmen, artists, or great thinkers, since environment, together with heredity, is a major factor in shaping our personality and skills.)

Cloning took a giant step closer to the realm of possibility in 1952, when Robert Briggs and Thomas King produced clones of frogs. They destroyed the nuclei of frogs' eggs with radiation and carefully

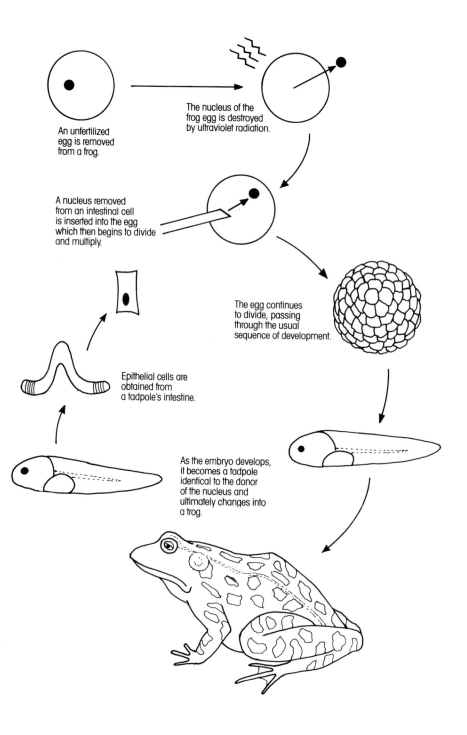

An unfertilized egg is removed from a frog.

The nucleus of the frog egg is destroyed by ultraviolet radiation.

A nucleus removed from an intestinal cell is inserted into the egg which then begins to divide and multiply.

The egg continues to divide, passing through the usual sequence of development.

Epithelial cells are obtained from a tadpole's intestine.

As the embryo develops, it becomes a tadpole identical to the donor of the nucleus and ultimately changes into a frog.

replaced them with nuclei taken from tadpoles of a different kind of frog. Some of these strange combinations developed into normal-looking frogs which were identical twins of the tadpoles from which the nuclei were taken. Similar experiments were later repeated by British biologist John Gurdon, using nuclei from skin and intestinal cells of well-developed tadpoles. In 1975 Gurdon's team managed to produce clones using skin cells from adult frogs, but some of the tadpoles were abnormal, and only one grew into an adult frog.

Attempts to clone mammals have had rather limited success so far. Mammalian egg cells are much smaller and more delicate than frog eggs, and they don't have the convenient habit of developing out in the open in pond water. J. Derek Bromhall of Oxford University has succeeded in cloning rabbits, using chemical fusion instead of delicate microsurgery. But he took the clones only up to an early stage of embryonic development.

In 1979 Karl Illmensee of the University of Geneva reported the successful cloning of mice. The Swiss researcher's approach was to inject a cell from a mouse embryo into a just-fertilized mouse egg and then remove the genetic material of the original sperm and egg cell. The nucleus from the donor cell directed the development of an embryo, which was transferred to the uterus

Cloning produces an "identical twin" from blueprints stored in the nuclei of body cells. The early cloning experiments were conducted on frogs; now variations of cloning have been produced with mammals as well.

of a mouse foster mother and grew into a normal mouse with the heredity of the donor cell, not of the fertilized egg.

Much research still lies ahead to determine whether such cloning techniques can be successfully repeated using cells from an adult mammal as the donor, rather than embryo cells, and to convert the process from the technical feat of a talented researcher to a routine laboratory procedure. But if this transition can be achieved, cloning opens up some intriguing possibilities. There may indeed be people someday who want to clone children to perpetuate their own image, or perhaps to clone spare body parts to replace a lost arm or leg or a failing organ. Cloning could have valuable applications in animal husbandry, yielding unlimited numbers of "carbon copies" of prize cattle, egg-laying hens, or champion racehorses.

Cloning may also help to preserve endangered species from the threat of extinction. The tiny flocks of whooping cranes, for example, could be increased by adding young birds cloned from cells of embryos or adults. Tissue samples of endangered creatures could be deep-frozen to provide insurance against a future time when more individuals might be needed to save the species from extinction. Indeed, scientists are even speculating on the possibility of recreating extinct species. Woolly mammoths—ancient relatives of today's elephants—have been found in the icy wastes of the far north, perfectly preserved in the Arctic ice. Could nuclei from cells of such mammoths, still containing the hereditary instructions for building a woolly mammoth, be inserted into elephant eggs, and grown in the body of an elephant foster mother?

Yale researcher Clement Markert is working on an interesting variation of cloning that may be valuable in animal husbandry. He takes an egg from a female mouse just after it has been fertilized, then removes the nucleus either from the egg or from the sperm. Chemical treatment with cytochalasin B prompts the remaining nucleus to duplicate its chromosomes, and then the egg begins to divide. When it is inserted into a mother mouse, it develops just like a normal embryo. The animal obtained is always a female, since only an XX chromosome combination could result from the doubling of one germ-cell nucleus. (If it were a Y sperm nucleus, the resulting YY cell could not survive.) The female produced is not a clone of either of her parents, since she shares just half the chromosome set of one of them. But when she matures, the eggs she produces all have exactly the same genes as her own set. If daughters are produced in the same way, removing the sperm nucleus, they will be clones of their mother. Markert speculates that his method could be used to breed lines of champion milk cows or egg-laying hens.

While some researchers have been breeding mice that have only one genetic parent, others have been working with another genetic oddity: mice whose heredity comes from *four* parents. They start with mice who have mated in the ordinary way. The fertilized eggs divide, again and again. When the embryos have reached the stage called a *blastocyst,* a tiny, partly hollow ball of cells, they are removed from the mother's body. Using tiny microsurgical tools, the researcher takes cells from

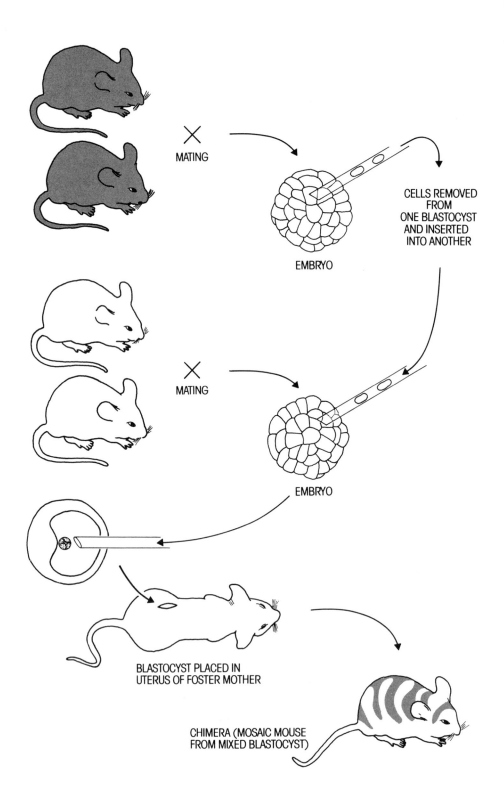

MATING

EMBRYO

CELLS REMOVED
FROM
ONE BLASTOCYST
AND INSERTED
INTO ANOTHER

MATING

EMBRYO

BLASTOCYST PLACED IN
UTERUS OF FOSTER MOTHER

CHIMERA (MOSAIC MOUSE
FROM MIXED BLASTOCYST)

one embryo and inserts them into the blastocyst of an embryo from a different mother. The new mixed blastocyst is placed into the uterus of a mouse and grows there into a normal-looking mouse, whose body contains cells from *both* of the original embryos. If one embryo came from a line of white-furred mice and the other from a line of dark-furred mice, the mixed offspring, called a *chimera,* will have patches of white and dark fur mingled in a mosaic pattern. Biochemical tests also show that some of the mouse's cells come from one set of parents and some from the other.

Chimeric mice are contributing to our knowledge of cancer. When embryos from a strain of mice that frequently develop cancer were combined with embryos from a strain that does not have this hereditary tendency, cancers in the chimeras developed only from the cells that came from the high-cancer strain. These studies support the theory that at least some cancers develop according to hereditary instructions carried within the body cells themselves.

A variation of the chimera studies is not only providing more information about cancer but may also be a valuable tool for studying various aspects of heredity. Cancer researcher Beatrice Mintz has produced mice that have a tumor as one of their parents! The tumor is a kind of cancer called teratoma,

When cells from one blastocyst are inserted into another, a chimera or mosaic offspring, expressing the heredity of all four parents, is produced.

which grows into an odd jumble of different kinds of tissues—almost like an embryo that got its instructions confused and didn't quite figure out how to make a proper baby. Teratoma cells grow rapidly in laboratory cultures. Mintz made chimeras by inserting teratoma cells from a culture into normal mouse blastocysts. Fur color and biochemical markers showed that the mice that grew from these altered embryos were real chimeras: Some of their cells had come from the original blastocyst and some from the teratoma. The portions of the mice that had developed from cancer cells were quite normal, and the mice proved to be no more likely to develop cancer than normal mice.

Being able to use cells in a culture as "parents" for mice opens up many intriguing possibilities. For example, Beatrice Mintz and Michael Dewey treated cultures of teratoma cells with chemical mutagens. Then they selected mutant cells lacking a particular enzyme, HPRT (hypoxanthine phosphoribosyltransferase), which is lacking in human patients with a genetic disease called Lesch-Nyhan syndrome. This is a sex-linked condition, the victims of which not only suffer from mental retardation and a spastic form of cerebral palsy, but also have a strange compulsion to hurt themselves by biting and scratching. The researchers inserted teratoma cells with the Lesch-Nyhan-like mutation into blastocysts and are studying the mixed-up mice obtained, in order to try to determine which tissues are the most important in producing or preventing the syndrome.

Another study used teratoma cells with a mutant gene causing anemia (a lack of enough red blood cells).

Chimeras with even a few normal cells in the blood-forming tissue did not develop anemia.

Studies like these could be used to study human diseases by actually introducing human genes into mice through treated teratoma cells. Researchers are also using chimeras with teratoma parents to study the DNA in the mitochondria, the cell's energy generators. This DNA is separate from the DNA in the chromosomes, but it can also determine some cell characteristics, and it can mutate. Mintz and her co-workers have already used teratoma cells to introduce a mutated mitochondrial gene, one that produces resistance to the antibiotic chloramphenicol, into mice. Until now, scientists have known very little about what role the mitochondrial DNA plays in heredity, development, and disease. Mouse chimeras may provide some answers.

Gene manipulators have also been making progress working with plants. Intriguing practical results are now starting to come in. If you have ever grown a whole carrot plant from a cut-off top or a potato plant from an "eye," you know that at least some plant cells retain the ability to grow into whole new plants. The offspring are "clones," sharing the heredity of the parent plant. So are plants grown from twigs or cuttings.

In the early 1960s, Cornell biologist Frederick Steward grew a carrot plant not from a whole cut-off top, but from a single carrot cell. Since then, whole plants have been grown from single cells in such diverse species as corn and pine trees. In an interesting variation, researchers at GTE Laboratories in Waltham, Massachusetts, are trying to get cells from fruits and vegetables to mature without growing whole plants. Tomato flowers snipped from young plants have been grown, on special nutrient media dosed with plant hormones, into red, ripe fruits. (The fruits are seedless, because the flowers were not pollinated.)

Some plant researchers are working on cell hybrids, produced by cell fusion. Getting plant cells to fuse poses somewhat different problems from those involved in fusing animal cells. Each plant cell is encased in a tough, rather rigid cell wall. To raise plant cells in cultures and to attempt fusion experiments, the researchers first had to devise suitable combinations of enzymes to dissolve away the cell wall, leaving "naked" protoplasts. Then methods had to be worked out to get the cells to fuse. Sendai virus does not work on plant cells. Eventually it was discovered that the chemical sodium nitrate makes naked plant cells crowd together; squeezing the cells together in a centrifuge makes some of them fuse. Later it was found that a chemical called polyethylene glycol (PEG) works even better, and also works on animal cells.

Peter Carlson and his research team at Brookhaven National Laboratory have fused cells of different kinds of tobacco and grown plants that are intermediates between the two parents. Agricultural researchers in Saskatchewan, Canada, are trying to combine the cells of much less closely related plants: soybeans and corn, soybeans and peas, soybeans and tobacco, and carrots and barley. So far they have gotten the cells to fuse and grow in cultures, but they have not yet been able to grow any plants from them. A hybrid of soybeans and corn, for example, could be very valuable, since it might combine the nutritional qualities of both parents, together with the ability of soybeans to form a partnership with nitrogen-fixing bacteria. Beans and other legumes can grow with much smaller additions of nitrogen fertilizers. If this ability could be transferred to other crop plants, agricultural production could be greatly increased.

Scientists are not sure if the goal of tranferring nitrogen-fixing abilities to wheat, corn, and other common crops is practical. Studies have shown, for example, that legumes devote a large percentage of their energy to their partnership with the bacteria that live in nodules in their roots. Producing similar partnerships for corn or wheat might cut down the yield of grain. But researchers are actively studying the genetics of nitrogen fixing and attempting to make it more efficient. They have isolated *nif* (nitrogen-fixing) genes and have succeeded in transferring working nif genes from one type of bacterium to another, using plasmids to carry the genes. Altering the genes of nitrogen-fixing soil bacteria so that they will be active at lower temperatures is another method being considered for increasing the supply of nitrogen available to plants.

Plant cell fusion experiments have recently come up with some peculiar results. The discovery that PEG works with both plant and animal cells gave researchers the idea of fusing plant cells with animal cells. Fuad Safwat of the University of Massachusetts in Boston, working with Edward Cocking in Nottingham, England, managed to fuse red blood cells from a chicken with protoplasts from yeasts. Later, the Brookhaven team fused tobacco cells with human cells. No attempt was made to grow the fused cells into any kind of creature. So far the "plantimals" are just laboratory curiosities.

Scientists have learned that they must proceed with caution in developing the new laboratory breeds of plants into new crops. Too great a reliance on a few genetic lines, even if they seem to have qualities far superior to the natural forms, might permit whole crops to be wiped out by new diseases. That is what happened when farmers began to rely on superior forms of hybrid corn. The widely planted hybrid turned out to be susceptible to a new type of corn blight that swept through the United States and wiped out 15 percent of the corn crop in 1970. A stand of identical cloned pine trees might meet a similar fate. But a careful mixture of plantings of several different superior stocks could avoid the problem while still profiting from the benefits provided by applied genetics.

In addition to increasing our supplies of food and raw materials, genetic engineering may help to solve other problems, too. In 1975 a research team at General Electric Company, headed by Ananda Chakrabarty, announced the development of oil-eating bacteria that could clean up oil spills. Recombinant DNA techniques were used to transfer genes from selected oil-eating strains of bacteria to a single form of *Pseudomonas*. The new hybrid is much more oil-hungry than the natural forms; it eats up more oil, faster. This new "bug" and other oil-eating strains that are being developed would serve as food for fish and other marine organisms after the oil spill is cleaned up.

Genetic engineering techniques can also be used to

improve microbes used for other purposes, such as the production of alcoholic beverages, cheeses, flavoring agents, hormones, and antibiotics. The production of "single-cell protein" substitutes for other foods is a growing industry. Microbes are also being used to decompose sewage and produce methane gas that can be used as a fuel.

Indeed, many industrialists are expecting recombinant DNA and other genetic engineering techniques to revolutionize the industry of the future. Oil companies, chemical manufacturers, and other major industrial firms are investing in "biotechnology" research in varied areas, from the production of foods and chemicals to the improvement of mining techniques. Bacterial enzymes, produced on a huge scale by specially programmed bacteria, can be used to regulate chemical reactions in factories. Such enzymes would provide a far finer control than is possible by purely chemical means. Some microorganisms naturally tend to take up a particular metal selectively from the food medium on which they grow. Using genetic engineering techniques, researchers can breed "superbugs" that specialize in concentrating particular metals such as gold or uranium. Such bacteria could then be used to treat ores, cheaply extracting valuable metals from them. In other projects, genetic engineers are breeding bacteria for producing alcohol, gums for use in recovering more oil from oil wells, flavoring and perfume agents, pesticides, lubricants, fertilizers, and drugs.

Researchers at G. D. Searle and Company are exploring the use of recombinant DNA techniques to produce a "universal" flu vaccine. The influenza virus has a tendency to mutate frequently, so that a vaccine that protected people against last year's flu epidemic might be ineffective against the strains that will be common next year. Drug companies have had to scramble continually to devise new vaccines to keep up with the changeable ways of the influenza virus. In 1979 Searle researchers transferred a gene coding for an influenza virus protein into bacteria, making it possible to produce large amounts of the virus protein. They hope that by making similar transfers for all the known flu viruses, they will be able to produce a vaccine against the virus proteins (rather than against a particular whole virus) that will work for any kind of influenza.

What about ourselves? Could human genes use some engineered improvement? The first applications of genetic engineering to humans will certainly be in the treatment of diseases—correcting nature's mistakes rather than trying to improve on nature's designs. But scientists—and science fiction writers—have speculated on some directions that gene modifications could take.

A knowledge of the on-off switches of the genes— how they work and how to manipulate them at will— would permit us to grow replacement organs for transplants from a person's own cells. There would be no worry about rejection of the transplants, for they would be recognized by the body as its own tissue. Replacement arms or legs or kidneys might even be grown right in place. The replacement of worn-out organs might take place gradually, cell by cell. If we could manipulate our genes to do these things, we could greatly increase our life span. Indeed, many researchers

Colonies of the oil-eating microbes developed by GE researchers can be seen in the culture dish on the left. The flasks on the right contain petroleum and water. The hybrid bacteria, added to one flask, have begun to digest the crude oil.

believe that aging is genetically controlled and is due to a gradual turning-off of working genes. If we learned to turn these genes back on again, we could stay youthful indefinitely.

The ability to reprogram our genes could also permit us to redesign people to live in unusual environments— under the sea or in the exotic atmospheres of other planets. A geneticist once happened to remark to a reporter that genetic engineering might permit us to produce "little green men" who could make their own food with the help of chlorophyll in their skins. The story was picked up in the news, and the researcher spent the next year trying to explain to indignant colleagues that he was only kidding. But some such variation might not be an entirely unreasonable possibility, sometime in the distant future.

Recent studies are indicating that certain types of behavior are strongly influenced by genes, and the genes for specific actions have even been mapped in fruit flies. Perhaps someday we may be able to modify genetically not only our bodies but also our personalities. Perhaps future genetic engineers will learn how to eliminate war.

These are some prospects of genetic engineering, in the near and distant future. Some will find them exciting. Some will find them terrifying. But the results of the genetics explosion in the years ahead are sure to be fascinating.

For Further Reading

Cooke, Robert. *Improving on Nature: The Brave New World of Genetic Engineering.* New York: Quadrangle/New York Times, 1977.

Goodfield, June. *Playing God: Genetic Engineering and the Manipulation of Life.* New York: Random House, 1977.

Halacy, D.S., Jr. *Genetic Revolution: Shaping Life for Tomorrow.* New York: Harper and Row, 1974.

Hyde, Margaret O. *The New Genetics.* New York: Franklin Watts, 1974.

Karp, Laurence E. *Genetic Engineering: Threat or Promise?* Chicago: Nelson Hall, 1976.

Lear, John. *Recombinant DNA: The Untold Story.* New York: Crown, 1978.

Rogers, Michael. *Biohazard.* New York: Alfred Knopf, 1977.

Wade, Nicholas. *The Ultimate Experiment: Man-Made Evolution.* New York: Walker, 1977.

Glossary

amniocentesis   a surgical procedure in which a sample of amniotic fluid is obtained through a hollow needle inserted into the uterus through the abdominal wall; the fluid sample is tested for chromosome changes and other indications of fetal abnormalities.

amplification of genes   the production of multiple copies of a plasmid in a bacterial cell, without division of the cell.

antibodies   specific proteins produced by the body's defense system in response to a foreign substance that has entered the body.

asexual   pertaining to a form of reproduction in which a single parent produces offspring.

bacteriophage   a type of virus that infects bacteria.

biological containment   the use of specially designed bacteria, incapable of living outside the laboratory, in recombinant DNA experiments.

cell fusion the combination of two independent cells (possibly from different species) into a single hybrid cell, usually after treatment with an agent such as Sendai virus. The hybrid cell retains some characteristics of each parent cell.

chimera an animal with a mixed genetic heritage, produced by mingling cells from two or more embryos.

chromosomes structures in the cell nucleus, each consisting of a long thread of DNA (carrying hereditary information), together with RNA and protein.

clone a cell culture, organism, or group of organisms all derived from a single parent cell and sharing the same heredity.

codon a sequence of three nucleotides in RNA, specifying a particular amino acid to be incorporated into a protein.

conjugation a kind of mating of microorganisms, in which hereditary information is exchanged.

cytochalasin B a drug that causes a cell to divide into a karyoplast (a nearly naked nucleus) and a cytoplast (a cell without a nucleus).

deoxyribonucleic acid the nucleic acid present in the chromosomes, in which the hereditary information is encoded. It consists of nucleotides, linked together in a double helix and each containing a sugar (deoxyribose), phosphate, and a nitrogen base (adenine, cytosine, guanine, or thymine).

DNA the abbreviation for deoxyribonucleic acid.

DNA ligase an enzyme that joins broken DNA chains together when the sticky ends are matched up.

dominant trait a hereditary characteristic that is expressed even if its gene is present on only one of the corresponding pair of chromosomes.

Down's syndrome a hereditary disorder characterized by mental retardation and various physical traits and due to trisomy of chromosome 21.

E. coli the abbreviation for *Escherichia coli,* a bacterium normally found in the human colon and commonly used in laboratory experiments.

gene the unit of heredity; a portion of the DNA on the

chromosome that directs the production of a specific protein and/or the expression of a specific trait.

genetic engineering   the practical application of genetics, such as the treatment of a disease by modifying the genes or introducing genes to direct the production of lacking biochemicals.

genetics   the science of heredity.

genome   the complete set of genes of an organism.

genotype   the genetic constitution of an individual.

germ cells   the reproductive cells (sperm, eggs, and the cells from which they are formed).

hemophilia   a hereditary disease in which the blood does not clot properly.

heterozygous   possessing two different genes for a particular trait, i.e., a mixed genotype.

homozygous   possessing two identical genes for a particular trait, i.e., a pure genotype.

Huntington's chorea   a hereditary degenerative disease, the symptoms of which usually do not appear until adulthood.

hybridoma   a form of cell fusion in which a cell producing a particular biochemical is combined with a hardy, fast-growing cancer cell to form a culture that synthesizes large quantities of the biochemical.

intervening sequences   segments of DNA in genes or RNA in messenger RNA that do not appear to carry a "message" of hereditary information and are not translated in protein synthesis.

messenger RNA   a form of RNA that carries the hereditary information encoded in a gene out from the nucleus into the cytoplasm, where it provides a pattern for protein synthesis in the ribosomes.

microcapsule   a tiny globule of fattylike material that can be used to enclose biochemicals of various kinds.

mitochondria   energy-producing structures inside the cell.

monoclonal antibodies   almost pure antibodies produced by hybridomas.

mutagenic   producing a change (mutation) in the genetic information of an organism or cell.

mutation a hereditary change in the genetic information of a cell or organism.

nondisjunction failure of the chromosomes to separate properly during the cell division forming the sperm or egg.

nucleic acid an organic compound consisting of nucleotides, chemically linked together in long chains. Various types of nucleic acids (DNA and RNAs) carry hereditary information and function in the production of proteins.

nucleoprotein an organic compound in which nucleic acids and proteins are combined.

nucleotide the structural unit of the nucleic acids, consisting of a sugar (ribose or deoxyribose), phosphate, and a nitrogen base (adenine, cytosine, guanine, thymine, or uracil).

phage a shortened name for bacteriophage.

phenotype the visible expression of a hereditary trait.

phenylketonuria a hereditary disease involving an inability to utilize the amino acid phenylalanine and leading to mental retardation.

physical containment a complex of protective measures, such as specially designed laboratories, to prevent the escape of microorganisms used in recombinant DNA research.

plasmid a small circular form of DNA found in some bacteria, which carries hereditary information outside the bacterial chromosome and can be transferred from one bacterium to another.

recessive trait a hereditary characteristic that is expressed only when the genes for it are present on both chromosomes of the corresponding pair (i.e., inherited from both parents), or when the gene determining the train is present on the X chromosome of a male.

recombinant DNA the transfer of genes from one species to another by cutting and splicing of DNA.

replication the reproduction of DNA.

restriction enzymes enzymes that cut a double strand of DNA, leaving sticky ends.

reverse transcriptase an enzyme that catalyzes the construction of

DNA according to an RNA template.

ribonucleic acid a nucleic acid consisting of nucleotides linked into a single helix and each containing the sugar ribose, phosphate, and a nitrogen base (adenine, cytosine, guanine, or uracil). Several types (messenger RNA, transfer RNA, and ribosomal RNA) function in protein synthesis.

RNA the abbreviation for ribonucleic acid.

ribosomal RNA the RNA found in the ribosomes, the structures where protein is synthesized.

Sendai virus a virus that is used (after treatment with ultraviolet light) to make cells fuse.

sex-linked recessive condition a hereditary trait that is transmitted by a gene on one of the sex chromosomes and thus is observed mainly in one sex. Sex-linked recessive traits carried on the X chromosome are much more common among males, since there they can be expressed even if they have been inherited from only one parent.

shotgunning a technique of recombinant DNA research in which the entire chromosome set of an organism is cut into fragments with restriction enzymes, and the fragments are inserted into plasmids for study.

sickle-cell anemia a hereditary disease in which some of the red blood cells have an abnormal form of hemoglobin and may collapse to a sickle shape, clumping together and clogging blood vessels.

Tay-Sachs disease a hereditary degenerative disease that causes death in early childhood.

template a pattern for nucleic acid synthesis.

transcription the production of an RNA copy of a DNA template.

transduction the transfer of hereditary information from one bacterium to another by means of infecting phages.

transfer RNA a form of RNA that recognizes a specific amino acid and carries it to the ribosomes for assembly into a protein.

translation the synthesis of a protein molecule according to the instructions coded in the nucleotides of messenger RNA.

translocation the transfer of a portion of a chromosome to another chromosome.

trisomy the presence in a cell of three of a particular type of chromosome instead of the normal two.

Index